Springer Proceedings in Physics

Volume 241

Indexed by Scopus

The series Springer Proceedings in Physics, founded in 1984, is devoted to timely reports of state-of-the-art developments in physics and related sciences. Typically based on material presented at conferences, workshops and similar scientific meetings, volumes published in this series will constitute a comprehensive up-to-date source of reference on a field or subfield of relevance in contemporary physics. Proposals must include the following:

- name, place and date of the scientific meeting
- a link to the committees (local organization, international advisors etc.)
- scientific description of the meeting
- list of invited/plenary speakers
- an estimate of the planned proceedings book parameters (number of pages/ articles, requested number of bulk copies, submission deadline).

More information about this series at http://www.springer.com/series/361

Michaela Kozlová · Jaroslav Nejdl
Editors

X-Ray Lasers 2018

Proceedings of the 16th International Conference on X-Ray Lasers

Springer

Editors
Michaela Kozlová
ELI Beamlines Project
Institute of Physics CAS
Praha 8, Czech Republic

Jaroslav Nejdl
ELI Beamlines Project
Institute of Physics CAS
Praha 8, Czech Republic

ISSN 0930-8989 ISSN 1867-4941 (electronic)
Springer Proceedings in Physics
ISBN 978-3-030-35452-7 ISBN 978-3-030-35453-4 (eBook)
https://doi.org/10.1007/978-3-030-35453-4

This Springer imprint is published by the registered company Springer Nature Switzerland AG
The registered company address is: Gewerbestrasse 11, 6330 Cham, Switzerland

Preface

This volume comprises papers presented at the 16th International Conference on X-Ray Lasers (ICXRL 2018) held at Café Louvre in Prague, Czech Republic, between October 7 and 12, 2018. This conference was the 16th in a series of biannual meetings that were held all over the world, starting with a meeting in Aussois, France, in 1986. The meeting was organized in the Czech Republic for the first time and gave, in oral and poster sessions, a broad review of the activities in the field of both coherent and incoherent X-ray sources and their scientific applications. The conference program was organized to promote closer contacts and better understanding between different subfields including plasma-based X-ray lasers, high-order harmonic generation, X-ray optics, imaging and metrology, and various applications.

The meeting brought 90 attendees from 10 countries in the world. The scientific schedule of the conference was supplemented by a social program, including a trip to Kutná Hora, a medieval town that was called the jewel and the treasury of the country. The conference dinner was held in the Nebozízek Restaurant located on the slope of Petřín Hill with marvelous views on the historical part of Prague.

For the first time, the Pierre Jaeglé prize for the best student contribution was awarded, which helped to strongly involve the young generation and boosted the intensity of discussions during the poster session. The prize was awarded to Malte Vassholz from the Institute for X-Ray Physics, University of Göttingen.

We would like to thank the members of the International Advisory Board of the Conference for their help in organizing the scientific program. There were also many local people, who contributed to the successful organization of the conference. We are not able to list all of them, but we would like to express our special thanks to Romana Kočová, Radka Alexa, Uddhab Chaulagain, Karel Boháček, and Ondřej Hort.

Finally, we are grateful to the conference attendees and contributors for their willingness to participate in the conference and finally to contribute to this book.

Prague, Czech Republic

Jaroslav Nejdl
Michaela Kozlová

Contents

Part I Plasma-Based X-Ray Lasers

1 **3D Chemical Mapping of Thin Films by Means of X-Ray Laser Microanalysis** 3
D. Bleiner, M. Trottmann, R. Müller, L. Rush, I. Kuznezov, A. Cabas-Vidani, Y. Romanyuk, A. Tiwari, G. R. Patzke, C. S. Menoni and J. J. Rocca

2 **The Ablation of Solid Targets Using a Capillary Discharge Laser** 11
G. J. Tallents, S. A. Wilson, J. Lolley, E. Solis Meza, E. Wagenaars, H. Bravo, C. S. Menoni and J. J. Rocca

3 **Hydrodynamic Modelling of Plasma-Based Soft X-ray Amplifiers** 17
E. Oliva

4 **The Use of Inhomogeneous Forming Lines in the Laser Power Supply System on a High-Current Z-Discharge Plasma** 23
N. Kalinin, V. Antonov, S. Vaganov and A. Kovalenko

5 **Enhanced Fluorescence from X-Ray Line Coincidence Pumping** 29
J. Nilsen, D. Burridge, L. M. R. Hobbs, D. Hoarty, P. Beiersdorfer, G. V. Brown, N. Hell, D. Panchenko, M. F. Gu, A. M. Saunders, H. A. Scott, P. Hatfield, M. P. Hill, L. Wilson, R. Charles, C. R. D. Brown and S. Rose

6 **Generation of Short Wavelength Radiation and Energetic Particles by Using PW-Class Lasers** 37
T. Kawachi, M. Nishikino, A. Sasaki, N. Hasegawa, M. Ishino, T.-H. Dinh, K. Mikami, T. Kitamura, H. Kiriyama, A. S. Pirozhkov, T. Th. Esirkepov, A. Sagisaka, M. Nishiuchi, H. Sakaki, M. Kando, A. Ya. Faenov, T. A. Pikuz, K. Kawase, R. Nagai, R. Hajima, S. V. Bulanov and K. Kondo

7 **Shaping of X-ray Pulses via Dynamical Control of Their Interaction with a Resonant Medium** 45
T. Akhmedzhanov, V. Antonov, X. Zhang, K. C. Han, E. Kuznetsova, I. Khairulin, Y. Radeonychev, M. Scully and O. Kocharovskaya

8 **Seeded Soft X-Ray Lasers: Towards OAM Beam Amplification and Quasi-Pi-pulse Regime** 53
O. Guilbaud, F. Sanson, A. K. Pandey, E. Papagiannouli, J. Demailly, O. Neveu, B. Lucas, E. Baynard, M. Pittman, D. Ros, F. Harms, G. Dovillaire, M. Richardson, Ph. Balcou and S. Kazamias

9 **EUV Laser Irradiation System with Intensity Monitor** 61
M. Ishino, T.-H. Dinh, N. Hasegawa, K. Sakaue, T. Higashiguchi, S. Ichimaru, M. Hatayama, M. Washio, M. Nishikino and T. Kawachi

10 **Li-like Aluminum Plasma Soft X-ray Laser by Means of Recombination Scheme** 67
S. Namba, N. Kubo, C. John, N. Hasegawa, M. Kishimoto, M. Nishikino and T. Kawachi

11 **Experimental Study of the Gain Dependence on Driving Laser Pulse Leading Edge Duration** 73
P. Zakopal, M. Krůs and M. Kozlová

Part II High-Order Harmonic Generation

12 **High-Charge XUV Vortex Beam: Generation and Hartmann Wavefront Sensor Characterization** 79
F. Sanson, A. K. Pandey, F. Harms, G. Dovillaire, E. Baynard, J. Demailly, O. Guilbaud, B. Lucas, O. Neveu, E. Papagionnouli, M. Pittman, D. Ros, M. Richardson, E. Johnson, W. Li, Ph. Balcou and S. Kazamias

13 **Ultrafast Inner-Shell Electron Excitation with High Energy Recollision Electron Driven by Mid-infrared Laser** 85
Z. Zeng, Y. Deng, Y. Zheng, G. Marcus and R. Li

14 User-Oriented High-Harmonic Source at ELI Beamlines 93
O. Finke, O. Hort, M. Albrecht, V. E. Nefedova, D. D. Mai, F. Giambruno, S. Reyne, L. Poletto, F. Frassetto, J. Gautier, S. Sebban, R. Antipenkov, F. Batysta, R. Boge, J. Naylon, Z. Hubka, T. Green, P. Bakule and J. Nejdl

15 Enhancement of Harmonics Generated in Modulated Indium Laser-Plasmas: Experiment and Theory 99
S. Y. Stremoukhov, R. A. Ganeev and A. V. Andreev

16 Imaging Plate Absolute Calibration in the XUV Wavelength Range 103
P. Gajdoš, M. Kozlová and M. Krůs

Part III X-Rays from Relativistic Electron Beams

17 Betatron X/γ-Ray Radiation from Wakefield-Accelerated Electrons Wiggling in Laser Fields 109
Y. F. Li, J. Feng, D. Z. Li, J. H. Tan, K. Huang, J. G. Wang, M. Z. Tao and L. M. Chen

18 LWFA-Driven Betatron Source for Plasma Physics Platform at ELI Beamlines 117
U. Chaulagain, K. Boháček, J. Vančura, M. Lamač, W. Yan, Y. Gu, M. Kozlová, K. Ta-Phuoc, S. A. Weber and J. Nejdl

19 Imaging Objects with Coded Apertures, Utilising a Laser Wakefield X-Ray Source 125
M. P. Selwood, R. Heathcote and C. D. Murphy

20 Progresses of a Hard X-Ray Split and Delay Line Unit for the MID Station at the European XFEL 131
W. Lu, B. Friedrich, T. Noll, K. Zhou, J. Hallmann, G. Ansaldi, T. Roth, S. Serkez, G. Geloni, A. Madsen and S. Eisebitt

21 Modeling Diamond Irradiated with a European XFEL Pulse 139
N. Medvedev

Part IV X-Ray Optics, Imaging and Metrology

22 Time-Resolved Near-Edge X-Ray Absorption Spectroscopy in the Water Window and Beyond Using Laboratory-Based Laser Plasma Sources 147
H. Stiel, D. Engel, L. Glöggler, A. Jonas, I. Mantouvalou and J. Tümmler

23 Ptychography and Single-Shot Nanoscale Imaging with Plasma-Based Laser Sources 155
F. Tuitje, M. Zürch, T. Helk, J. Gautier, F. Tissandier, J.-P. Goddet, E. Oliva, A. Guggenmos, U. Kleineberg, H. Stiel, S. Sebban and C. Spielmann

24 Pico–Femto–Attosecond Metrology for Advanced and XUV Photon Sources 163
T. Sinyakova, C. Bomme, D. Cubaynes, C. Bourassin-Bouchet, A. Gharbi, G. Garcia, A. K. Pandey, I. Papagiannouli, O. Guilbaud and A. Klisnick

25 High Spectral and Spatial Resolution Soft X-ray/XUV VLS Spectrographs 169
E. N. Ragozin, A. O. Kolesnikov, A. S. Pirozhkov, P. V. Sasorov, A. N. Shatokhin and E. A. Vishnyakov

26 Spectral and Polarization Properties of VUV-Mirrors for Experiments at a HHG Beamline 175
S. Richter, S. Espinoza and J. Andreasson

27 On Inverse Problem for Propagation of Waves from Inclined Surfaces 181
R. M. Feshchenko

28 Flux and Energy Density of a Collapsing Electromagnetic Pulse in the Free Space 187
I. A. Artyukov, A. V. Vinogradov, N. V. Dyachkov and R. M. Feshchenko

Author Index 197

Contributors

T. Akhmedzhanov Department of Physics and Astronomy, Texas A&M University, College Station, TX, USA

M. Albrecht ELI Beamlines, Institute of Physics ASCR, Prague, Czech Republic; FNSPE, Czech Technical University in Prague, Prague, Czech Republic

J. Andreasson ELI Beamlines/Fyzikální ústav AV ČR, v.v.i., Dolní Břežany, Czech Republic;
Chalmers tekniska högskola, Institutionen för fysik, Göteborg, Sweden

A. V. Andreev Faculty of Physics, Lomonosov Moscow State University, Moscow, Russia

G. Ansaldi European X-Ray Free-Electron Laser Facility, Schenefeld, Germany

R. Antipenkov ELI Beamlines, Institute of Physics ASCR, Prague, Czech Republic

V. Antonov Peter the Great St. Petersburg Polytechnic University, St. Petersburg, Russia

V. Antonov Institute of Applied Physics, Russian Academy of Sciences, Nizhny Novgorod, Russia;
Prokhorov General Physics Institute, Russian Academy of Sciences, Moscow, Russia

I. A. Artyukov P.N. Lebedev Physical Institute RAS, Moscow, Russia

P. Bakule ELI Beamlines, Institute of Physics ASCR, Prague, Czech Republic

Ph. Balcou Centre Laser Intense et Applications, Université de Bordeaux (UMR-CNRS-CEA 5107), Talence, France

F. Batysta ELI Beamlines, Institute of Physics ASCR, Prague, Czech Republic; FNSPE, Czech Technical University in Prague, Prague, Czech Republic

E. Baynard LASERIX, Centre Laser Université Paris Sud, FR2764, Université Paris-Saclay (UMR-CNRS 8578), Orsay, France

P. Beiersdorfer Lawrence Livermore National Laboratory, Livermore, CA, USA

D. Bleiner Swiss Federal Laboratories for Materials Science and Technology (Empa), Dübendorf, Switzerland;
Department of Chemistry, University of Zurich, Zurich, Switzerland

R. Boge ELI Beamlines, Institute of Physics ASCR, Prague, Czech Republic

K. Boháček ELI Beamlines, Institute of Physics ASCR, Prague, Czech Republic; FNSPE, Czech Technical University in Prague, Prague, Czech Republic

C. Bomme CEA-Saclay, LIDYL, Saclay, France

C. Bourassin-Bouchet Institut of Optique, LCF, Palaiseau, France

H. Bravo Department of Electrical and Computer Engineering, Center for Extreme Ultraviolet Science and Technology, Colorado State University, Fort Collins, USA

C. R. D. Brown AWE Plc, Aldermaston, UK

G. V. Brown Lawrence Livermore National Laboratory, Livermore, CA, USA

S. V. Bulanov Kansai Photon Science Institute (KPSI), National Institutes for Quantum and Radiological Science and Technology (QST), Kyoto, Japan;
ERT/High Field Initiative (HIFI), ELI-Beamlines, Dolní Břežany, Czechia

D. Burridge AWE Plc, Aldermaston, UK

A. Cabas-Vidani Swiss Federal Laboratories for Materials Science and Technology (Empa), Dübendorf, Switzerland

R. Charles AWE Plc, Aldermaston, UK

U. Chaulagain ELI Beamlines, Institute of Physics ASCR, Prague, Czech Republic

L. M. Chen Beijing National Laboratory of Condensed Matter Physics, Institute of Physics, CAS, Beijing, China;
School of Physical Sciences, University of Chinese Academy of Sciences, Beijing, China;
IFSA Collaborative Innovation Center, Department of Physics and Astronomy, Shanghai Jiao Tong University, Shanghai, China

D. Cubaynes CNRS, ISMO, Paris-Sud University, Orsay, France

J. Demailly Laboratoire de Physique des Gaz et des Plasmas, Université Paris-Saclay (UMR-CNRS 8578), Orsay, France

Y. Deng SwissFEL, Paul Scherrer Institut (PSI), Villigen, Switzerland

T.-H. Dinh Kansai Photon Science Institute (KPSI), National Institutes for Quantum and Radiological Science and Technology (QST), Kyoto, Japan

G. Dovillaire Imagine Optic, Orsay, France

N. V. Dyachkov P.N. Lebedev Physical Institute RAS, Moscow, Russia;
Moscow Institute of Physics and Technology (State University), Dolgoprudnyi, Moscow, Russia

S. Eisebitt Max Born Institute, Berlin, Germany;
Insitut für Optik und Atomare Physik, Technische Universität Berlin, Berlin, Germany

D. Engel Max-Born-Institute, Berlin, Germany

T. Th. Esirkepov Kansai Photon Science Institute (KPSI), National Institutes for Quantum and Radiological Science and Technology (QST), Kyoto, Japan

S. Espinoza ELI Beamlines/Fyzikální ústav AV ČR, v.v.i., Dolní Břežany, Czech Republic

A. Ya. Faenov Institute for Open and Transdisciplinary Research Initiatives, Osaka University, Osaka, Japan;
Joint Institute for High Temperature of the Russian Academy of Science, Moscow, Russia

J. Feng Beijing National Laboratory of Condensed Matter Physics, Institute of Physics, CAS, Beijing, China

R. M. Feshchenko P.N. Lebedev Physical Institute RAS, Moscow, Russia

O. Finke ELI Beamlines, Institute of Physics ASCR, Prague, Czech Republic;
FNSPE, Czech Technical University in Prague, Prague, Czech Republic

F. Frassetto Institute of Photonics and Nanotechnologies, University of Padua, Padua, Italy

B. Friedrich Max Born Institute, Berlin, Germany

P. Gajdoš IPP CAS, Prague 8, Czech Republic

R. A. Ganeev Faculty of Physics, Voronezh State University, Voronezh, Russia;
The Guo China-US Photonics Laboratory, State Key Laboratory of Applied Optics, Changchun Institute of Optics, Fine Mechanics and Physics, Chinese Academy of Sciences, Changchun, China

G. Garcia Synchrotron SOLEIL, Saint-Aubin, France

J. Gautier Laboratoire d'optique appliquée, ENSTA-ParisTech, Palaiseau, France;
ELI Beamlines, Institute of Physics ASCR, Prague, Czech Republic

G. Geloni European X-Ray Free-Electron Laser Facility, Schenefeld, Germany

A. Gharbi CNRS, ISMO, Paris-Sud University, Orsay, France

F. Giambruno ARDOP Industrie, Cité de la Photonique, Pessac, France

L. Glöggler Berlin Laboratory for Innovative X-Ray Technologies (BLiX), Berlin, Germany;
Institut für Optik und Atomare Physik, Technische Universität Berlin, Berlin, Germany

J.-P. Goddet Laboratoire d'optique appliquée, ENSTA-ParisTech, Palaiseau, France

T. Green ELI Beamlines, Institute of Physics ASCR, Prague, Czech Republic

Y. Gu ELI Beamlines, Institute of Physics ASCR, Prague, Czech Republic;
Institute of Plasma Physics ASCR, Prague, Czech Republic

M. F. Gu University of California, Berkeley, CA, USA

A. Guggenmos Department of Chemistry, University of California at Berkeley, Berkeley, USA;
Department for Physics, Ludwig-Maximilian-University Munich, Garching, Germany

O. Guilbaud LPGP, CNRS, Paris-Sud University, Orsay, France;
Laboratoire de Physique des Gaz et des Plasmas, Université Paris-Saclay (UMR-CNRS 8578), Orsay, France

R. Hajima Takasaki Advanced Radiation Research Institute, National Institutes for Quantum and Radiological Science and Technology (QST), Ibaraki, Japan

J. Hallmann European X-Ray Free-Electron Laser Facility, Schenefeld, Germany

K. C. Han Department of Physics and Astronomy, Texas A&M University, College Station, TX, USA

F. Harms Imagine Optic, Orsay, France

N. Hasegawa Kansai Photon Science Institute (KPSI), National Institutes for Quantum and Radiological Science and Technology (QST), Kizugawa, Kyoto, Japan

M. Hatayama NTT Advanced Technology, Atsugi, Kanagawa, Japan

P. Hatfield Imperial College, London, UK

R. Heathcote Central Laser Facility, STFC Rutherford Appleton Laboratory, Didcot, Oxon, UK

T. Helk Institute for Optics and Quantum Electronics, Abbe Center of Photonics, University of Jena, Jena, Germany;
Helmholtz Institute Jena, Jena, Germany

N. Hell Lawrence Livermore National Laboratory, Livermore, CA, USA

T. Higashiguchi Utsunomiya University, Utsunomiya, Tochigi, Japan

M. P. Hill AWE Plc, Aldermaston, UK

D. Hoarty AWE Plc, Aldermaston, UK

L. M. R. Hobbs AWE Plc, Aldermaston, UK

O. Hort ELI Beamlines, Institute of Physics ASCR, Prague, Czech Republic

K. Huang Beijing National Laboratory of Condensed Matter Physics, Institute of Physics, CAS, Beijing, China

Z. Hubka ELI Beamlines, Institute of Physics ASCR, Prague, Czech Republic; FNSPE, Czech Technical University in Prague, Prague, Czech Republic

S. Ichimaru NTT Advanced Technology, Atsugi, Kanagawa, Japan

M. Ishino Kansai Photon Science Institute (KPSI), National Institutes for Quantum and Radiological Science and Technology (QST), Kyoto, Japan

C. John Graduate School of Engineering, Hiroshima University, Higashihiroshima, Hiroshima, Japan

E. Johnson Department of Electrical and Computer Engineering, Clemson University, Clemson, SC, USA

A. Jonas Berlin Laboratory for Innovative X-Ray Technologies (BLiX), Berlin, Germany;
Institut für Optik und Atomare Physik, Technische Universität Berlin, Berlin, Germany

N. Kalinin Joffe Physical-Technical Institution Russian Academy of Science, St. Petersburg, Russia

M. Kando Kansai Photon Science Institute (KPSI), National Institutes for Quantum and Radiological Science and Technology (QST), Kyoto, Japan

T. Kawachi Kansai Photon Science Institute, National Institute for Quantum and Radiological Science and Technology, Kizu, Kyoto, Japan;
National Institutes for Quantum and Radiological Science and Technology, Kizugawa, Kyoto, Japan

K. Kawase Takasaki Advanced Radiation Research Institute, National Institutes for Quantum and Radiological Science and Technology (QST), Ibaraki, Japan

S. Kazamias Laboratoire de Physique des Gaz et des Plasmas, Université Paris-Saclay (UMR-CNRS 8578), Orsay, France

I. Khairulin Institute of Applied Physics, Russian Academy of Sciences, Nizhny Novgorod, Russia

H. Kiriyama Kansai Photon Science Institute (KPSI), National Institutes for Quantum and Radiological Science and Technology (QST), Kyoto, Japan

M. Kishimoto Kansai Photon Science Institute, National Institute for Quantum and Radiological Science and Technology, Kizu, Kyoto, Japan

T. Kitamura Kansai Photon Science Institute (KPSI), National Institutes for Quantum and Radiological Science and Technology (QST), Kyoto, Japan

U. Kleineberg Department for Physics, Ludwig-Maximilian-University Munich, Garching, Germany

A. Klisnick CNRS, ISMO, Paris-Sud University, Orsay, France

O. Kocharovskaya Department of Physics and Astronomy, Texas A&M University, College Station, TX, USA

A. O. Kolesnikov P.N. Lebedev Physical Institute, Russian Academy of Sciences, Moscow, Russia;
Moscow Institute of Physics and Technology (National Research University), Dolgoprudnyi, Moscow Region, Russia

K. Kondo Kansai Photon Science Institute (KPSI), National Institutes for Quantum and Radiological Science and Technology (QST), Kyoto, Japan

A. Kovalenko Joffe Physical-Technical Institution Russian Academy of Science, St. Petersburg, Russia

M. Kozlová Institute of Plasma Physics CAS, Prague, Czech Republic;
Institute of Physics CAS, Prague, Czech Republic;
ELI Beamlines, Institute of Physics ASCR, Prague, Czech Republic

M. Krůs Institute of Plasma Physics CAS, Prague, Czech Republic

N. Kubo Graduate School of Engineering, Hiroshima University, Higashihiroshima, Hiroshima, Japan

E. Kuznetsova Department of Physics and Astronomy, Texas A&M University, College Station, TX, USA

I. Kuznezov Department of Electrical and Computer Engineering, NSF Center for Extreme Ultraviolet Science and Technology, Colorado State University, Fort Collins, Colorado, USA

M. Lamač ELI Beamlines, Institute of Physics ASCR, Prague, Czech Republic;
Charles University, Prague, Czech Republic

D. Z. Li Institute of High Energy Physics, CAS, Beijing, China

R. Li State Key Laboratory of High Field Laser Physics, Shanghai Institute of Optics and Fine Mechanics, Chinese Academy of Sciences, Shanghai, China

W. Li Department of Electrical and Computer Engineering, Clemson University, Clemson, SC, USA

Y. F. Li Beijing National Laboratory of Condensed Matter Physics, Institute of Physics, CAS, Beijing, China

J. Lolley Department of Physics, York Plasma Institute, University of York, York, UK

W. Lu European X-Ray Free-Electron Laser Facility, Schenefeld, Germany

B. Lucas Laboratoire de Physique des Gaz et des Plasmas, Université Paris-Saclay (UMR-CNRS 8578), Orsay, France

A. Madsen European X-Ray Free-Electron Laser Facility, Schenefeld, Germany

D. D. Mai ELI Beamlines, Institute of Physics ASCR, Prague, Czech Republic

I. Mantouvalou Berlin Laboratory for Innovative X-Ray Technologies (BLiX), Berlin, Germany;
Institut für Optik und Atomare Physik, Technische Universität Berlin, Berlin, Germany

G. Marcus Department of Applied Physics, Benin School of Engineering and Computer Science, Hebrew University of Jerusalem, Jerusalem, Israel

N. Medvedev Institute of Physics, Czech Academy of Sciences, Prague, Czechia; Institute of Plasma Physics, Czech Academy of Sciences, Prague, Czechia

C. S. Menoni Department of Electrical and Computer Engineering, Center for Extreme Ultraviolet Science and Technology, Colorado State University, Fort Collins, USA

K. Mikami Kansai Photon Science Institute (KPSI), National Institutes for Quantum and Radiological Science and Technology (QST), Kyoto, Japan

R. Müller Department of Chemistry, University of Zurich, Zurich, Switzerland

C. D. Murphy Department of Physics, York Plasma Institute, University of York, York, UK

R. Nagai Takasaki Advanced Radiation Research Institute, National Institutes for Quantum and Radiological Science and Technology (QST), Ibaraki, Japan

S. Namba Graduate School of Engineering, Hiroshima University, Higashihiroshima, Hiroshima, Japan

J. Naylon ELI Beamlines, Institute of Physics ASCR, Prague, Czech Republic

V. E. Nefedova ELI Beamlines, Institute of Physics ASCR, Prague, Czech Republic;
FNSPE, Czech Technical University in Prague, Prague, Czech Republic

J. Nejdl ELI Beamlines, Institute of Physics ASCR, Prague, Czech Republic

O. Neveu Laboratoire de Physique des Gaz et des Plasmas, Université Paris-Saclay (UMR-CNRS 8578), Orsay, France

J. Nilsen Lawrence Livermore National Laboratory, Livermore, CA, USA

M. Nishikino Kansai Photon Science Institute (KPSI), National Institutes for Quantum and Radiological Science and Technology (QST), Kyoto, Japan

M. Nishiuchi Kansai Photon Science Institute (KPSI), National Institutes for Quantum and Radiological Science and Technology (QST), Kyoto, Japan

T. Noll Max Born Institute, Berlin, Germany

E. Oliva Departamento de Ingeníeria Energética, ETSI Industriales, Universidad Politécnica de Madrid, Instituto de Fusión Nuclear, Universidad Politecnica de Madrid, Madrid, Spain

D. Panchenko Lawrence Livermore National Laboratory, Livermore, CA, USA

A. K. Pandey Laboratoire de Physique des Gaz et des Plasmas, Université Paris-Saclay (UMR-CNRS 8578), Orsay, France

E. Papagionnouli Laboratoire de Physique des Gaz et des Plasmas, Université Paris-Saclay (UMR-CNRS 8578), Orsay, France

I. Papagiannouli LPGP, CNRS, Paris-Sud University, Orsay, France

G. R. Patzke Department of Chemistry, University of Zurich, Zurich, Switzerland

T. A. Pikuz Institute for Open and Transdisciplinary Research Initiatives, Osaka University, Osaka, Japan;
Joint Institute for High Temperature of the Russian Academy of Science, Moscow, Russia

A. S. Pirozhkov Kansai Photon Science Institute (KPSI), National Institutes for Quantum and Radiological Science and Technology (QST), Kizugawa, Kyoto, Japan

M. Pittman LASERIX, Centre Laser Université Paris Sud, FR2764, Université Paris-Saclay (UMR-CNRS 8578), Orsay, France

L. Poletto Institute of Photonics and Nanotechnologies, University of Padua, Padua, Italy

Y. Radeonychev Institute of Applied Physics, Russian Academy of Sciences, Nizhny Novgorod, Russia

E. N. Ragozin P.N. Lebedev Physical Institute, Russian Academy of Sciences, Moscow, Russia

S. Reyne ARDOP Industrie, Cité de la Photonique, Pessac, France

M. Richardson University of Central Florida, Orlando, FL, USA;
Centre Laser Intense et Applications, Université de Bordeaux (UMR-CNRS-CEA 5107), Talence, France

S. Richter ELI Beamlines/Fyzikální ústav AV ČR, v.v.i., Dolní Břežany, Czech Republic

J. J. Rocca Department of Electrical and Computer Engineering, NSF Center for Extreme Ultraviolet Science and Technology, Colorado State University, Fort Collins, Colorado, USA

Y. Romanyuk Swiss Federal Laboratories for Materials Science and Technology (Empa), Dübendorf, Switzerland

D. Ros Laboratoire de Physique des Gaz et des Plasmas, Université Paris-Saclay (UMR-CNRS 8578), Orsay, France

S. Rose Imperial College, London, UK

T. Roth ESRF – The European Synchrotron, Grenoble, France

L. Rush Department of Electrical and Computer Engineering, NSF Center for Extreme Ultraviolet Science and Technology, Colorado State University, Fort Collins, Colorado, USA

A. Sagisaka Kansai Photon Science Institute (KPSI), National Institutes for Quantum and Radiological Science and Technology (QST), Kyoto, Japan

H. Sakaki Kansai Photon Science Institute (KPSI), National Institutes for Quantum and Radiological Science and Technology (QST), Kyoto, Japan

K. Sakaue Waseda University, Shinjuku, Tokyo, Japan

F. Sanson Laboratoire de Physique des Gaz et des Plasmas, Université Paris-Saclay (UMR-CNRS 8578), Orsay, France;
Amplitude Laser Group, Scientific Business Unit-Lisses Operations, Lisses, France

A. Sasaki Kansai Photon Science Institute (KPSI), National Institutes for Quantum and Radiological Science and Technology (QST), Kyoto, Japan

P. V. Sasorov Institute of Physics ASCR, v.v.i. (FZU), ELI-Beamlines Project, Prague, Czech Republic

A. M. Saunders Lawrence Livermore National Laboratory, Livermore, CA, USA

H. A. Scott Lawrence Livermore National Laboratory, Livermore, CA, USA

M. Scully Department of Physics and Astronomy, Texas A&M University, College Station, TX, USA

S. Sebban Laboratoire d'optique appliquée, ENSTA-ParisTech, Palaiseau, France;
ELI Beamlines, Institute of Physics ASCR, Prague, Czech Republic

M. P. Selwood Department of Physics, York Plasma Institute, University of York, York, UK

S. Serkez European X-Ray Free-Electron Laser Facility, Schenefeld, Germany

A. N. Shatokhin P.N. Lebedev Physical Institute, Russian Academy of Sciences, Moscow, Russia;
Moscow Institute of Physics and Technology (National Research University), Dolgoprudnyi, Moscow Region, Russia

T. Sinyakova CNRS, ISMO, Paris-Sud University, Orsay, France

E. Solis Meza Department of Physics, York Plasma Institute, University of York, York, UK

C. Spielmann Institute for Optics and Quantum Electronics, Abbe Center of Photonics, University of Jena, Jena, Germany;
Helmholtz Institute Jena, Jena, Germany

H. Stiel Max-Born Institute Berlin, Berlin, Germany;
Berlin Laboratory for Innovative X-Ray Technologies (BLiX), Berlin, Germany

S. Y. Stremoukhov National Research Centre "Kurchatov Institute", Moscow, Russia;
Faculty of Physics, Lomonosov Moscow State University, Moscow, Russia

G. J. Tallents Department of Physics, York Plasma Institute, University of York, York, UK

J. H. Tan Beijing National Laboratory of Condensed Matter Physics, Institute of Physics, CAS, Beijing, China

M. Z. Tao Beijing National Laboratory of Condensed Matter Physics, Institute of Physics, CAS, Beijing, China

K. Ta-Phuoc Laboratoire d'Optique Appliquée, ENSTA, CNRS UMR7639, École Polytechnique, Palaiseau, France

F. Tissandier Laboratoire d'optique appliquée, ENSTA-ParisTech, Palaiseau, France

A. Tiwari Swiss Federal Laboratories for Materials Science and Technology (Empa), Dübendorf, Switzerland

M. Trottmann Swiss Federal Laboratories for Materials Science and Technology (Empa), Dübendorf, Switzerland

F. Tuitje Institute for Optics and Quantum Electronics, Abbe Center of Photonics, University of Jena, Jena, Germany;
Helmholtz Institute Jena, Jena, Germany

J. Tümmler Max-Born-Institute, Berlin, Germany;
Berlin Laboratory for Innovative X-Ray Technologies (BLiX), Berlin, Germany

S. Vaganov Joffe Physical-Technical Institution Russian Academy of Science, St. Petersburg, Russia

J. Vančura ELI Beamlines, Institute of Physics ASCR, Prague, Czech Republic; FNSPE, Czech Technical University in Prague, Prague, Czech Republic

A. V. Vinogradov P.N. Lebedev Physical Institute RAS, Moscow, Russia

E. A. Vishnyakov P.N. Lebedev Physical Institute, Russian Academy of Sciences, Moscow, Russia

E. Wagenaars Department of Physics, York Plasma Institute, University of York, York, UK

J. G. Wang Beijing National Laboratory of Condensed Matter Physics, Institute of Physics, CAS, Beijing, China

M. Washio Waseda University, Shinjuku, Tokyo, Japan

S. A. Weber ELI Beamlines, Institute of Physics ASCR, Prague, Czech Republic

L. Wilson AWE Plc, Aldermaston, UK

S. A. Wilson Department of Physics, York Plasma Institute, University of York, York, UK

W. Yan ELI Beamlines, Institute of Physics ASCR, Prague, Czech Republic

P. Zakopal Institute of Plasma Physics CAS, Prague, Czech Republic

Z. Zeng State Key Laboratory of High Field Laser Physics, Shanghai Institute of Optics and Fine Mechanics, Chinese Academy of Sciences, Shanghai, China

X. Zhang Department of Physics and Astronomy, Texas A&M University, College Station, TX, USA

Y. Zheng State Key Laboratory of High Field Laser Physics, Shanghai Institute of Optics and Fine Mechanics, Chinese Academy of Sciences, Shanghai, China

K. Zhou Shanghai Institute of Applied Physics, Chinese Academy of Sciences, Shanghai, China;
University of Chinese Academy of Sciences, Beijing, China

M. Zürch Institute for Optics and Quantum Electronics, Abbe Center of Photonics, University of Jena, Jena, Germany;
Helmholtz Institute Jena, Jena, Germany;
Department of Chemistry, University of California at Berkeley, Berkeley, USA;
Fritz Haber Institute of the Max Planck Society, Berlin, Germany

Part I
Plasma-Based X-Ray Lasers

Chapter 1
3D Chemical Mapping of Thin Films by Means of X-Ray Laser Microanalysis

D. Bleiner, M. Trottmann, R. Müller, L. Rush, I. Kuznezov, A. Cabas-Vidani, Y. Romanyuk, A. Tiwari, G. R. Patzke, C. S. Menoni and J. J. Rocca

Abstract A capillary discharge X-ray laser is a tabletop coherent source, which has matured for application in microanalysis. Soft X-ray pulses were delivered onto functional materials to induce photo-ablation in the nanoscale. Such samplings were measured inline by means of mass spectrometry, and thus perform depth profile analysis. This paper discusses the complementarity or this technique to contending microanalytical techniques, to have a direct methodology for space resolved *in situ* thin film characterization.

1.1 Introduction

Space-resolved solid *microanalysis* is a fundamental process in the context of materials design and development. The ability to control locally the chemical and physical quality of products of synthesis gets increasingly more and more challenging as the features get smaller. *Smaller* is intended in different dimensions, i.e., in *concentration*, needing techniques able to detect the trace-contaminant signals resolved from the matrix, and in *length-scale*, in order to catch any local heterogeneity or impurity. This is a crucial validation procedure, also for identifying impurities in the processing tools, e.g., chambers, manipulators, *etc.*, as an uncontrolled source of contamination and malfunction of the prepared functional thin films [19].

D. Bleiner (✉) · M. Trottmann · A. Cabas-Vidani · Y. Romanyuk · A. Tiwari
Swiss Federal Laboratories for Materials Science and Technology (Empa), Überlandstrasse 129, CH-8600 Dübendorf, Switzerland
e-mail: davide.bleiner@empa.ch

D. Bleiner · R. Müller · G. R. Patzke
Department of Chemistry, University of Zurich, Winterthurerstrasse 190, CH-8057 Zurich, Switzerland

L. Rush · I. Kuznezov · C. S. Menoni · J. J. Rocca
Department of Electrical and Computer Engineering, NSF Center for Extreme Ultraviolet Science and Technology, Colorado State University, Fort Collins, Colorado 80523, USA

M. Kozlová and J. Nejdl (eds.), *X-Ray Lasers 2018*, Springer Proceedings in Physics 241, https://doi.org/10.1007/978-3-030-35453-4_1

For instance, the structure and purity of multilayers are crucial for their functionality as short-wavelength optics [1], photovoltaic units [2], and so forth. Microanalysis, based on focused pulses, allows collecting detailed chemical information in a material, in a vectorial mode [3–8]. It is beyond the scope of this paper to provide an extensive review of the various techniques, and the interested reader is recommended to refer to the available literature [9].

Figure 1.1 summarizes the capabilities of a selection of state-of-the-art techniques. The different depth ranges are indicated with a specific nomenclature, which has been adopted in the scientific community. One observes that the domain of "*thin films*" is not *directly* accessible with any of the available techniques. However, sometimes one could utilize dynamic sputtering methods, whose basic capability is at shorter length-scales, and by extending the sputtering time, also cover the desired range.

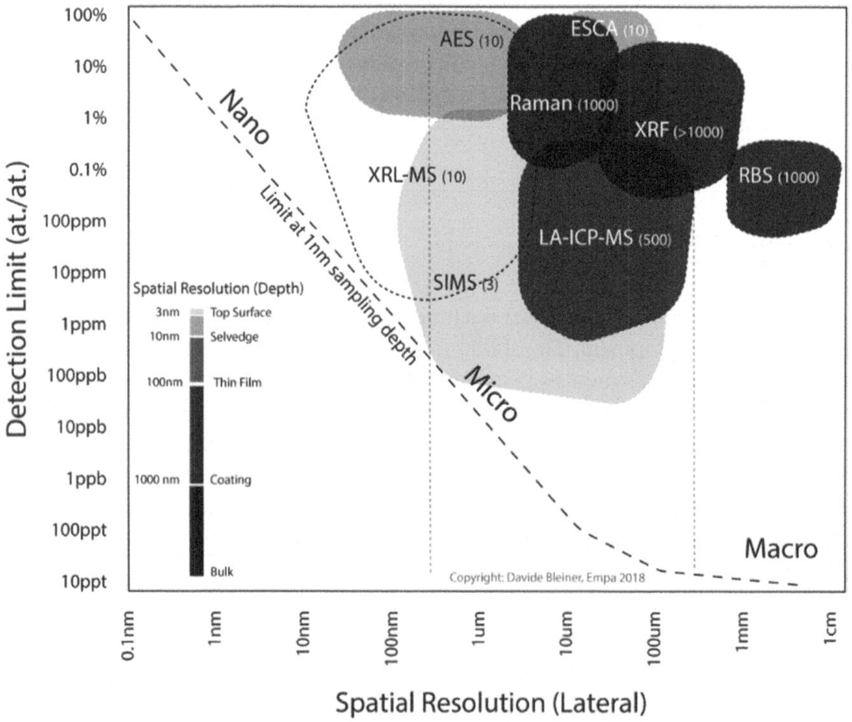

Fig. 1.1 Detection limit as a function of spatial resolution for a selection of microanalytical techniques. The depth resolution is given in the scale of grays. It is to note that the techniques are far from the theoretical limit, due to the low sample utilization efficiency. Further, no technique can directly access the thin-film length-scale, which is a potential for the X-ray laser microanalysis. In parenthesis is the depth resolution in nm. *Legend* AES Auger Electron Spectroscopy, ESCA Electron Spectroscopy for Chemical Analysis, LA-ICP-MS Laser Ablation Inductively Coupled Plasma Mass Spectrometry, RBS Rutherford Backscattering Spectroscopy, SIMS Secondary Ion Mass Spectrometry, XRF X-Ray Fluorescence, XRL-MS X-ray Laser Mass Spectrometry

Nevertheless, such an extended process may have two major drawbacks, namely: (i) altering the material in the spot's neighborhood, (ii) generating sequential and time-consuming data acquisition protocols that are inefficient or incompatible with *operando* mode.

X-ray laser nano-ablation (XRL-MS) is thus promising to fill the state of the art's gap. A reason for technology *readiness* [10] delay is the lack of accessible sources. Latest generation X-ray lasers are in fact based on complex and large-scale accelerator facilities, i.e., fourth generation beamlines ("X-ray Free-Electron Lasers", or XFEL). The latter are unfortunately hardly accessible, though detain their niche impact for advanced *proof-of-principle* research. Important activities such as daily R&D in academic or industrial groups, practical training of students, high-risk research with *high failure/high gain* connotation, are few notable examples of "*no go*" activities at XFEL. Tabletop sources, therefore, regardless of the more modest investment compared to the multi-millions for accelerators, and simpler engineering, are being investigated at a few pioneering groups [11, 12].

This brief introductory discussion has highlighted the need for 24/7 accessible X-ray sources [11–15] e.g., for direct thin film microanalysis, with trace impurity sensitivity. Sensitivity and space resolution are mutually exclusive figures-of-merit [4], because the drastic reduction of probed mass degrades the counting statistics, and with that the detection sensitivity.

The aim of this paper is to discuss capabilities of plasma-driven tabletop X-ray lasers for direct thin film microanalysis, as a lateral and depth-wise chemical mapping (3D). Fundamental advantages of the methodology for analytical application are discussed. Two materials science applications are shown as highlighted from dedicated experimental studies.

1.2 Materials and Methods

The experimental setup utilized was discussed in detail elsewhere [16, 17]. The capillary discharge X-ray-laser ($\lambda = 46.9$ nm or 26 eV) generated pulses of about 10 μJ with duration (FWHM) of about 2 ns and a pulse to pulse energy fluctuation of 10%. Grazing incidence optics were used to collimate the X-ray-laser and focus it on the sample using a zone plate lens. The ablated ions were accelerated to 1.5 keV energy due to the potential difference between the sample and the zone plate, which were 2.13 mm apart. The zone plate had a central opening of 50 μm in order to allow the ablated ions to be injected into the time of flight mass spectrometry. The entire system occupies an optical table space of 0.6×3 m. Typically an X-ray-irradiation pattern of 1000 points arranged as a $10 \times 10 \times 10$ grid on each sample was carried out. The aerial spot size was 1 μm while the center-to-center distance between the adjacent points was 2 μm. 2D map stacks were generated by interpolation of the resulting data grids of spectra.

1.3 Results and Discussions

1.3.1 Comparison to State-of-the-Art Methods

Figure 1.1 compares the capabilities of state-of-the-art microanalytical techniques. One observes that a diagonal, a fundamental limit is present. Such a limit derives from the effective counting statistics that is associated with a certain sampling volume. One can qualitatively understand that the reduction of the spatial resolution degrades the sensitivity with a technique-specific correlation coefficient. Most technique works in the micro-range.

1.3.2 Advantages of the X-Ray Laser for Chemical Mapping

1.3.2.1 Ablation Threshold

The laser ablation process requires that an excess of energy is delivered to the solid material, in the form of a burst of photons. Short pulses have the advantage that the peak power may easily exceed the minimum requirement to initiate the ablative process, thanks to multi-photon effects, even if the energy of a single photon may be incapable of ablation. This explains why femtosecond pulses are often utilized. On the other hand, the high photon energy of X-ray pulses may make think that the ablation process is virtually threshold free, because a single photon suffices to trigger ablation. Studies [18 and references therein] have, however, shown that even if up to one order of magnitude lower, even short-wavelength, e.g., soft X-rays, exhibit a threshold. Indeed, one can explain this experimental observation considering that solid ablation at short wavelength is not energy-limited, but mass-limited. In fact, ablation is a collective mass disruption process, and a certain threshold above a few photon irradiations must be achieved to observe a significant mass removal.

1.3.2.2 Access to Hard Bonds

Typical chemical bond energies are in the range 1–10 eV and ionization energy of elements as high as 24.6 eV, while the photon energy of commercial laser systems is as high as 6.4 eV (ArF excimer). It is clear that the resonant photo-dissociation of X-ray pulses of 25 eV and above are ideal to ablate any matrix without significant thermal effects. Of course, the cross section must be taken into account, and so harder X-ray pulses may result as inefficient as longer wavelength with respect to ablation efficiency. A heat-free (resonant) ablation process is important for stoichiometric analysis as well as for micro-structuring.

1.3.3 Selected Thin Film Applications

1.3.3.1 Stability of CoNCN Catalytic Thin Film

The development of renewable energy sources based on the conversion of water into oxygen and hydrogen as a storable primary fuel depends critically on the construction of efficient electrocatalysts. Informed catalyst design requires full control over the structural features of such catalysts before and after processing.

Given that the catalyst performance depends on the composition and stability of the top thin film, they must be characterized with spectroscopic tools enabling nanoscale resolution. In Müller et al. [17] we have demonstrated that X-ray laser ablation offers the required flexibility to test rapidly in 3D (surface and depth-wise) the characteristics of a nanostructured layer. The electrode surface with electrocatalyst cobalt carbodiimide (CoNCN) was characterized. The comparison of 3D maps of the X-ray Laser mass spectra of pre- and post-catalytic electrodes (Fig. 1.2) revealed substantial changes and pointed towards catalyst layer detachment during electrocatalysis. Such insights are essential for the development of stable electrocatalysts in many forefront areas of contemporary energy and materials research.

Figure 1.2 shows a selection of chemical compounds identified by a 3D hyperspectral matrix. A few chemicals are key indicators of the materials' functional elements. The element *F* is contained in the particle binder Nafion, while *C* is the main element in the substrate that when directly ablated builds a dimer *C*2. Along with the distribution of the functional elements Co and *N*, these observations support the statement

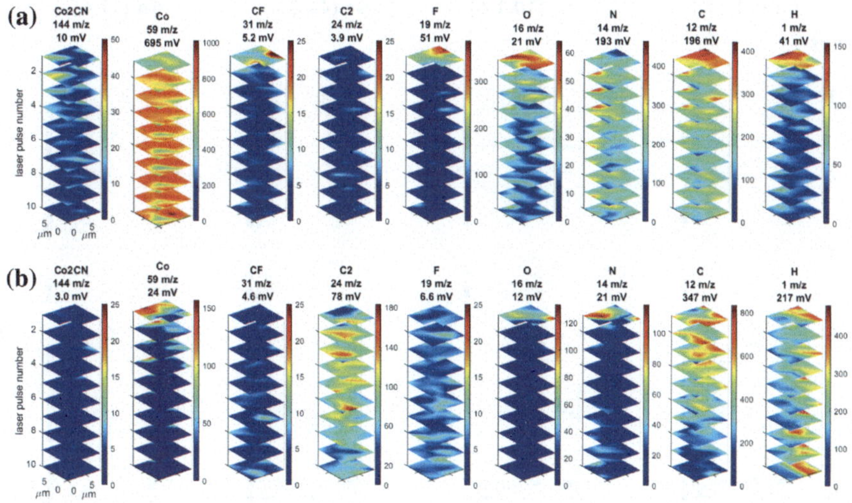

Fig. 1.2 3D chemical mapping of CoNCN catalytic thin film for water-oxidation. **a** Pre-catalytic, **b** Post-catalytic

above about the lack of catalytic stability. It should be said that a number of alternative analytical methods were utilized, but none of these was able to offer both wide surface and depth-wise thin film investigation. The strength of X-ray laser ablation mass spectrometry is that one can generate full 3D hyperspectral data.

1.3.3.2 Structure and Alloying of Kesterite Photovoltaic Thin Film

The performance-enhancement effect of alkali-doping is well-known for chalcogenide thin-film solar cells, based on Cu(In, Ga)Se_2 (CIGS) or $Cu_2ZnSn(S, Se)_4$ (CZTSSe–kesterite) absorbers. In contrast to heavier alkali elements, lithium was found to alloy with the kesterite phase [2] leading to the solid solution $(Li_xCu_{1-x})_2ZnSn(S, Se)_4$. The latter opened a way of tuning the semiconductor bandgap by changing the Li/(Li + Cu) ratio.

The detailed characterization of photovoltaic thin films, i.e., with space-resolved capability, is important to validate all nominal manufacturing characteristics. In particular, the ability to map the chemical composition in 3D is important to a compositional control of the tuning quality.

Figure 1.3 shows a stack of 2D mappings at increasing depth. Each depth layer refers to a specific laser short, from the first one (at surface) to the tenth one (at depth). One observes the granular structure given by the polycrystallinity. The granular structure modulates the chemical response in a correlated way between the various elements. The main elements of the CZTS samples are shown. The anticorrelation

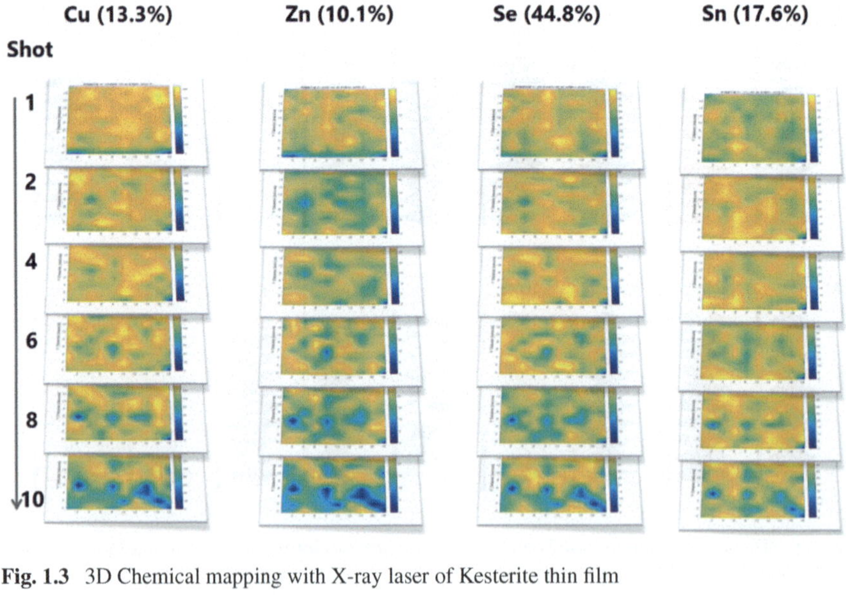

Fig. 1.3 3D Chemical mapping with X-ray laser of Kesterite thin film

between Li and Cu (not shown due to editorial limits), related to substitution, was observed.

The mass removal per single shot was determined by knowing the thickness of the individual layers. A cross section prepared with a focused ion beam trench and electron microscopy imaging provided information on the structure. Visualized discontinuities were also identified in the depth profiling analysis with the X-ray Laser Mass Spectrometry, shown in Fig. 1.3. One can thus estimate an ablation yield of 15 nm/shot for the CZTS (shallow) layer, while the buried Mo layer was ablated at 40 nm/shot. The significant difference in ablation yield is to be attributed to matrix effects, i.e., the different absorption cross section and mass density.

1.4 Conclusions

Laser ablation mass spectrometry is a well-established microanalytical technique. The use of a soft X-ray laser at 46.9 nm permits to drastically reduce the spatial resolution as well as to ablate any kind of material. The observation that the ablation yield is matrix-dependent recommends future work to develop a fundamental understanding and strategies to correct for, in order to obtain accurate quantitative depth calibration data.

References

1. D. Bleiner, S. Yulin, J. Martynczuk, M. Ruiz-Lopez, Y. Arbelo, J.E. Balmer, D. Günther, Actinic damage of Y/Mo multilayer optics in a table-top plasma-driven x-ray laser. Appl. Opt. (22), 4894–4902 (2014)
2. A. Cabas-Vidani, S.G. Haass, C. Andres, R. Caballero, R. Figi, C. Schreiner, J.A. Márquez, C. Hages, T. Unold, D. Bleiner et al., High-efficiency $(Li_xCu_{1-x})_2ZnSn(S,Se)_4$ kesterite solar cells with lithium alloying. Adv. Energy Mater. (2018)
3. L. Armelao, D. Bleiner, V. Di Noto, S. Gross, C. Sada, U. Schubert, E. Tondello, H. Vonmont, A. Zattin, Ion-, photoelectron- and laser-assisted analytical investigation of nano-structured mixed HfO_2–SiO_2 and ZrO_2–SiO_2 thin films. Appl. Surf. Sci. **249**(1–4), 277–294 (2005)
4. D. Bleiner, P. Lienemann, A. Ulrich, H. Vonmont, A. Wichser, Spatially resolved quantitative profiling of compositionally graded perovskite layers using laser ablation-inductively coupled plasma mass spectrometry. J. Anal. At. Spectrom. **18**(9), 1146–1153 (2003)
5. D. Bleiner, F. Belloni, D. Doria, A. Lorusso, V. Nassisi, Overcoming pulse mixing and signal tailing in laser ablation inductively coupled plasma mass spectrometry depth profiling. J. Anal. At. Spectrom. **20**(12), 1337–1343 (2005)
6. D. Bleiner, P. Lienemann, H. Vonmont, Laser-induced particulate as carrier of analytical information in LA-ICPMS direct solid microanalysis. Talanta **65**(5), 1286–1294 (2005)
7. D. Bleiner, Mathematical modelling of laser-induced particulate formation in direct solid microanalysis. Spectrochim. Acta—Part B At. Spectrosc. **60**(1), 49–64 (2005)
8. D. Bleiner, P. Gasser, Structural features of laser ablation particulate from Si target, as revealed by focused ion beam technology. Appl. Phys. A: Mater. Sci. Process. **79**(4–6), 1019–1022 (2004)

9. H. Bubert, H. Jenett (eds.), *Surface and Thin Film Analysis: A Compendium of Principles, Instrumentation and Applications* (Wiley-VCH), ISBN 978-3527304585
10. J.A. Mitchell, *Measuring the Maturity of a Technology* (Sandia Rep. SAND2007-6733, 2007)
11. D. Bleiner, Coherent short-wavelength plasma radiation for lab-scale nano-inspection tools, in *X-Ray Lasers 2012,* eds. by S. Sebban et al., Springer Proceedings in Physics, vol. 147
12. D. Bleiner, A. Michette et al. (eds.), *Short Wavelength Laboratory Sources: Principles and Practices* (Royal Society of Chemistry, 2015)
13. M. Ruiz-Lopez, D. Bleiner, Implementing the plasma-lasing potential for tabletop nano-imaging. Appl. Phys. B: Lasers Opt. **115**(3), 311–324 (2014)
14. D. Bleiner, Y. Arbelo-Pena, L. Masoudnia, M. Ruiz-Lopez, Table-top x-ray lasers using a plasma gain-medium: Limits and potentials. Physica Scripta, T162. Article number 014050 (2014)
15. D. Bleiner, F. Staub, V. Guzenko, Y. Ekinci, J.E. Balmer, Evaluation of lab-scale EUV microscopy using a table-top laser source. Opt. Commun. **284**(19), 4577–4583 (2011)
16. I. Kuznetsov, J. Filevich, F. Dong, M. Woolston, W. Chao, E.H. Anderson, E.R. Bernstein, D.C. Crick, J.J. Rocca, C.S. Menoni, Three-dimensional nanoscale molecular imaging by extreme ultraviolet laser ablation mass spectrometry. Nat. Commun. **6**, 6944 (2015)
17. R. Müller, I. Kuznetsov, Y. Arbelo, M. Trottmann, C.S. Menoni, J.J. Rocca, G.R. Patzke, D. Bleiner, Depth-profiling microanalysis of CoNCN water-oxidation catalyst using a $\lambda = 46.9$ nm plasma laser for nano-ionization mass spectrometry. Anal. Chem. **90**(15), 9234–9240 (2018)
18. L. Juha et al., Short-Wavelength ablation of molecular solids: pulse duration and wavelength effects. J. Microlith. Microfab. Microsyst. 033007 (2005)
19. F. Belloni, D. Doria, A. Lorusso, V. Nassisi, L. Torrisi, L. Calcagnile, G. Quarta, D. Bleiner, D. Manno, Characterization of ablation plasma ion implantation. Nucl. Instrum. Methods Phys. Res., Sect. B: Beam Interact. Mater. At. **240**, 36–39 (2005)

Chapter 2
The Ablation of Solid Targets Using a Capillary Discharge Laser

G. J. Tallents, S. A. Wilson, J. Lolley, E. Solis Meza, E. Wagenaars, H. Bravo, C. S. Menoni and J. J. Rocca

Abstract Capillary discharge lasers with output at wavelength 46.9 nm offer a route to laboratory investigations of warm dense matter and may have an impact in the micro-ablation of solid samples of relevance to microelectromechanical systems. The commissioning and operation of a capillary discharge laser with 50 μJ output per pulse is presented alongside initial results of target ablation obtained by focusing the laser output. Studies have been undertaken relevant to the operation of the capillary laser. The capillary laser discharges a current pulse of up to 40 kAmps through an argon plasma, which is initially singly ionized via radio-frequency heating.

2.1 Introduction

Lasing at wavelength 46.9 nm in the extreme ultraviolet (EUV) has been achieved using capillary discharge plasmas [1–3], with several applications of the laser output now being developed (for example, see [4, 5]). Amplified spontaneous emission occurs along typically 0.2 m of argon plasma ionized and excited by sequences of radio-frequency heating and electrical discharges. The physics of a capillary discharge plasma is similar to more conventional Z-pinch designs, but with a solid capillary acting as a physical guide for the electrical current [6]. Electron temperatures in the pinch plasma can approach 80 eV for times of several nanoseconds and enable the production of population inversions in Ne-like argon leading to lasing

G. J. Tallents (✉) · S. A. Wilson · J. Lolley · E. Solis Meza · E. Wagenaars
Department of Physics, York Plasma Institute, University of York,
York YO105DD, UK
e-mail: greg.tallents@york.ac.uk

H. Bravo · C. S. Menoni · J. J. Rocca
Department of Electrical and Computer Engineering,
Center for Extreme Ultraviolet Science and Technology,
Colorado State University, Fort Collins, USA

M. Kozlová and J. Nejdl (eds.), *X-Ray Lasers 2018*, Springer Proceedings in Physics 241, https://doi.org/10.1007/978-3-030-35453-4_2

at 46.9 nm [1, 2]. We have shown [7] that visible spectral emission from the argon plasma medium of the capillary laser can be used to determine the electron temperature of the plasma created initially by using the Rayleigh–Jeans expression for black-body emission [8]. In this paper, we describe experiments where the capillary laser output was focused onto solid targets and present initial results of the target ablation.

2.2 Experiment

The capillary discharge laser output propagated in vacuum a distance of 1.35 m onto a normalized spherical mirror which focused the laser output to a small diameter on small plane targets (see Fig. 2.1). The target dimensions were small so that only approximately 10% of the laser energy was blocked by the back of the targets. Targets of solid gold, copper, aluminium or poly(methylacrylate) (PMA) were ablated by the laser and the craters examined using an atomic force microscope (see Fig. 2.2). Using spherical mirrors to focus the laser beam in the direction back towards the laser output significantly reduces the aberration of the spherical mirrors. Gold mirrors with a reflectivity at 46.9 nm of approximately 8–10% and silicon–scandium multilayer mirrors with reflectivities of approximately 50% were used enabling fluences on target of up to 400 J cm^{-2}. Crater diameters of minimum size ≈3 μm were produced. In a single shot, crater depths were measured at up to 1 μm.

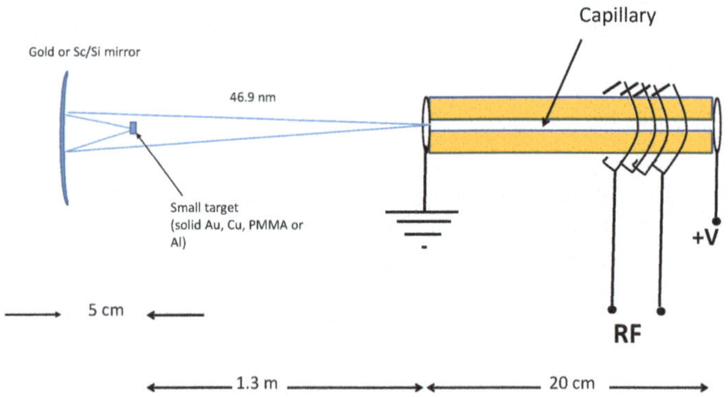

Fig. 2.1 A schematic of the argon-filled (0.3 Torr) capillary laser with radio-frequency (RF) heating coil and electrodes for high-voltage direct current discharge down the 3 mm diameter bore of the capillary. The spherical mirror focuses the laser output on the targets as illustrated

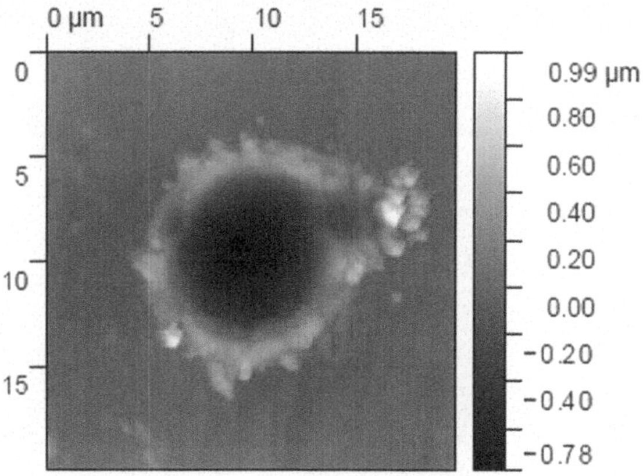

Fig. 2.2 An atomic force microscope image of the crater formed in a solid gold target by focusing the EUV capillary laser

2.3 Discussion

The laser output has an annular profile in the far field with a peak to peak angular range of approximately 5 mrad (see Fig. 2.3). We employed spherically focusing mirrors with 100 mm radius of curvature giving a focal length of 50 mm. The annular beam can be focused to a diameter d_f determined by the range of angles to the laser axis as given by the Abbe criterion. We have

$$d_f \approx \frac{\lambda}{\phi_{\max} - \phi_{\min}}$$

where $\phi_{\max}$ is the maximum angle of focused rays to the laser axis, $\phi_{\min}$ is the minimum angle of focused rays to the laser axis and $\lambda = 46.9$ nm is the laser wavelength. We estimate that $\phi_{\max} \approx 0.09$ radians and $\phi_{\min} \approx 0.05$ radians which gives a diffraction-limited focal diameter $d_f \approx 1.2$ μm. Evaluation of the aberrations via a 'circle of least confusion' calculation produced with rays focused at $\phi_{\max} \approx 0.09$ radians and $\phi_{\min} \approx 0.05$ radians with spherical focusing mirrors of radius of curvature 100 mm indicates that the diameter of the focused laser beam due to aberrations is approximately 2 μm. Our minimum measured crater diameters are ≈ 3 μm which suggests that the craters may be larger than the achieved focal diameters due to lateral transport of energy.

Figure 2.3 illustrates some of the issues in recording EUV laser output. Filter materials are often of poor quality due to small changes in thickness causing big changes in EUV transmission. The 0.8 μm thick aluminium used in Fig. 2.3 introduces structure in the images of the beam and is more transparent at high photon energies than at the laser photon energy. For example, at wavelength 46.9 nm, 0.8 μm thick aluminium transmits a fraction 0.15 so that X-rays of energy greater than 500 eV (wavelengths < 2.5 nm) are more readily transmitted.

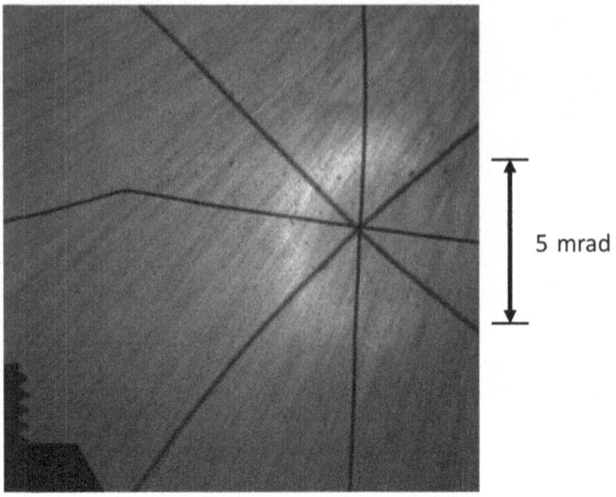

Fig. 2.3 An image recorded on a back-thinned CCD camera of the laser output. The image is obtained in the far field after ≈45° reflection from a spherical gold mirror and has a horizontal scale reduced by $\approx 1/\sqrt{2}$ compared to the marked vertical angular divergence scale. Alignment cross-wires and a mirror post shadowing the radiation are evident. There is evidence for background harder X-ray flux from the capillary plasma. The 46.9 nm output is represented by the annular shape around the confluence of the cross-wires. The striations arise from structure on the surface of the gold mirror and structure in a 0.8 μm aluminium filter covering the CCD

Acknowledgements We acknowledge financial support through the Centre for Doctoral Training in Fusion Energy (EPSRC grant EP/K504178/1).

References

1. C.D. Machchietto, B.R. Benware, J.J. Rocca, Generation of millijoule-level soft-x-ray laser pulses at a 4-Hz repetition rate in a highly saturated tabletop capillary discharge amplifier. Opt. Lett. **24**, 1115 (1999)
2. J.J. Rocca, Table-top soft X-ray lasers. Rev. Sci. Instrum. **70**, 3799 (1999)
3. S. Heinbuch, M. Grisham, D. Martz, J.J. Rocca, Demonstration of a desk-top size high repetition rate soft x-ray laser. Opt. Express **13**, 4050 (2005). https://doi.org/10.1364/OPEX.13.004050
4. A.K. Rossall, V. Aslanyan, G.J. Tallents, I. Kuznetsov, J.J. Rocca, C.S. Menoni, Ablation of submicrometer holes using an extreme ultraviolet laser. Phys. Rev. Appl. **3**, 064013 (2015). https://doi.org/10.1103/PhysRevApplied.3.064013
5. V. Aslanyan, I. Kuznetsov, H. Bravo, M.R. Woolston, A.K. Rossall, C.S. Menoni, J.J. Rocca, G.J. Tallents, Ablation and transmission of thin solid targets irradiated by intense extreme ultraviolet laser radiation. APL Photonics **1**, 066101 (2016). https://doi.org/10.1063/1.4953669
6. V.N. Shlyaptsev, J. Dunn, S.J. Moon et al., Modeling of capillary discharge plasma for x-ray lasers, XUV lithography and other applications, in *IEEE 5th International Conference on Dense Z-Pinches*, ed. by J. David et al. (2002). https://doi.org/10.1063/1.1531364

7. T. Page, S.A. Wilson, E. Branson, E. Wagenaars, G.J. Tallents, Plasma temperature measurements using black-body radiation from spectral lines emitted by a capillary discharge. JQSRT **220**, 1–4 (2018). https://doi.org/10.1016/j.jqsrt.2018.09.005
8. G.J. Tallents, An introduction to the atomic and radiation physics of plasmas (Cambridge University Press, Cambridge, 2018). https://doi.org/10.1017/9781108303538

Chapter 3
Hydrodynamic Modelling of Plasma-Based Soft X-ray Amplifiers

E. Oliva

Abstract In this article, we will show recent hydrodynamic simulations of quasi-steady state (QSS) plasma-based amplifiers of soft X-rays. These simulations are being performed with the 2D radiative-hydrodynamics code ARWEN. Selected results among the simulations performed will be presented, showing the inadequacy of some materials compared to others (Molybdenum and Zinc) for QSS scheme, mainly due to energy loses via radiative transfer. The importance of radiative transfer will be further assessed. Finally, ongoing results on the optimization of Titanium QSS plasma-based soft X-ray laser will be presented.

3.1 Introduction

The performance of plasma-based soft X-ray lasers strongly depends on the plasma conditions. Transient collisional excitation (TCE) [10] and quasi-steady state [15] (QSS) schemes rely on the absorption of laser beams in a preformed plasma. Optical field ionization scheme [8] requires the propagation of an intense infrared pulse throughout several millimetres of gas or even a hydrodynamic preformed plasma waveguide [6, 7]. In all these cases, the hydrodynamic behaviour of the plasma plays a fundamental role. Several hydrodynamic codes have been used to model these amplifiers, like EHYBRID [14] and ARWEN [11]. With the latter code it was demonstrated that bidimensional effects must be taken into account to correctly model

E. Oliva (✉)
Instituto de Fusión Nuclear "Guillermo Velarde", Universidad Politécnica de Madrid, Madrid, Spain
e-mail: eduardo.oliva@upm.es

Departamento de Ingeniería Energética, E.T.S.I. Industriales, Universidad Politécnica de Madrid, Madrid, Spain

Institute of Plasma Physics, AS CR, Za Slovankou 3, Prague 8, 18200 Prague, Czech Republic

M. Kozlová and J. Nejdl (eds.), *X-Ray Lasers 2018*, Springer Proceedings in Physics 241, https://doi.org/10.1007/978-3-030-35453-4_3

energy loses [2] and hydrodynamic expansion [13]. Recently, the code ARWEN was used to model the hydrodynamic generation of a plasma waveguide required for high- density OFI amplifiers [12].

In this paper we will show ongoing hydrodynamic modellization of QSS amplifiers that can be created and pumped at PALS facility (Prague Asterix Laser System). We carry out these simulations with the 2D radiative-hydrodynamics code ARWEN [11]. This code uses the Adaptive Mesh Refinement technique [1] which allows to study big domains while maintaining enough resolution in specific regions without increasing computational resources. Radiative transfer is modelled with a discrete-ordinate radiation transport method [11]. The equation of state (EOS) [4] and opacities [5] are computed and given to ARWEN as datafiles. This code has been extensively applied in fields other than plasma-based soft X-ray lasers, as inertial confinement fusion [17] and laboratory astrophysics [3, 9, 16].

3.2 Hydrodynamic Modelling of QSS Amplifiers

In order to achieve optimal conditions for amplification of soft X-rays, it is necessary to test several target materials (Molybdenum, Zinc, Silver, Titanium, etc…). Thus, it is necessary to feed the hydrodynamic code with the corresponding equation of state (EOS) and opacities for each material. Radiation transfer plays an important role. The radiated energy cools the central part of the plasma while heating parts of the target that were not lighted by the laser. This effect creates lateral wings of plasma that collide with the main plasma plume, modifying the posterior hydrodynamic expansion.

Figure 3.1 shows the comparison of a Molybdenum (left) and a Zinc (right) plasma. Both plasmas are created with a laser prepulse of 300 ps duration and an energy of 2.3 J arriving at $t = 1.5$ ns and a main pulse of 350 ps duration and an energy of 700 J arriving at $t = 11.5$ ns. Both pulses have a focal spot of 130 m. The images are taken at $t = 12$ ns. In the case of Molybdenum, radiative transfer cools the central part

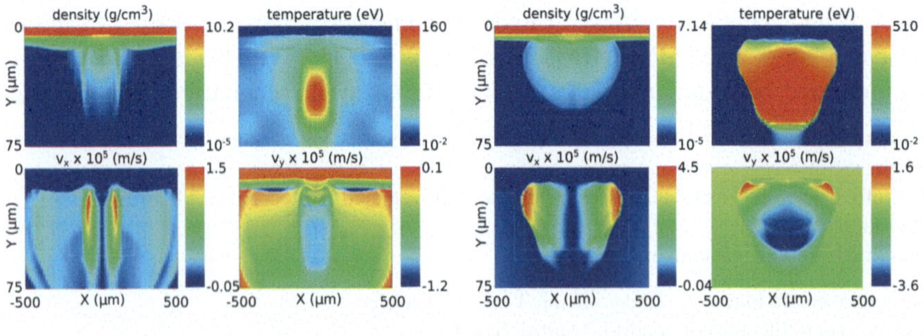

Fig. 3.1 (left) Electron density, temperature, horizontal and vertical expansion velocities for a Molybdenum plasma as given by our hydrodynamic model. (right) The same for a Zinc plasma

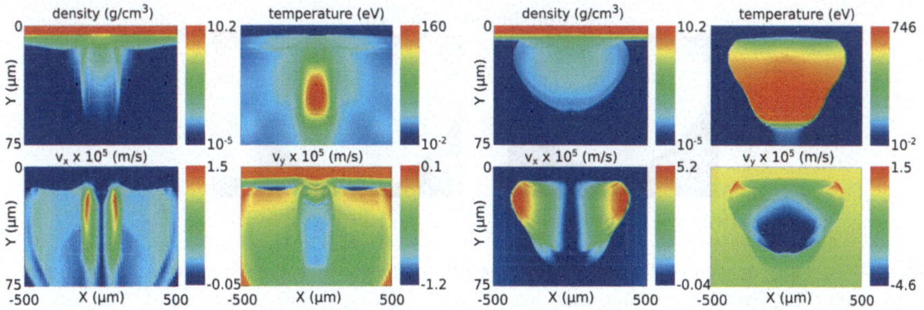

Fig. 3.2 (left) Electron density, temperature, horizontal and vertical expansion velocities for a Molybdenum plasma as given by our hydrodynamic code when radiative transfer is activated (left) The same with radiative transfer deactivated (right)

of the plasma while creating side plasma wings in zones of the target far from the laser focal spot. The resulting hydrodynamic collision between the main plasma and these wings creates a complicated profile, with jets (as shown in the density map). This cooling is not so important in Zinc plasmas, maintaining a relatively warm and homogeneous central zone well suited for the creation of a population inversion and posterior amplification of soft x-rays.

The importance of this radiative transfer is highlighted in Fig. 3.2, where the hydrodynamic expansion of the same Molybdenum plasma is shown with (left) and without (right) radiative transfer.

Without the cooling effect of radiative transfer, the central part of the plasma maintains a warm and homogeneous profile, as when Zinc was used. In addition to this, the lateral expansion of the plasma is no more hindered by the aforementioned lateral wings resulting in a bigger region fit for amplification when no radiation loses take place. Indeed, Zinc is a well-known QSS soft X-ray amplifier [15], while Molybdenum behaves better on other lasing schemes where the electron fluid is out of equilibrium. Thus, the role of radiative transfer and of the material chosen has been assessed with hydrodynamic simulations. As it was aforementioned, other materials and combinations of prepulse-main pulse have been tested in order to find the best conditions for QSS lasing.

As an example, Fig. 3.3 shows the electron density, as given by our hydrodynamic code, of a Titanium plasma 7.4 ns after the arrival of an infrared ($\lambda = 1.315$ μm) laser pulse, with 300 ps duration focused in a 700 μm × 4 cm line. The energy of the pulse is 32.4 J in the case depicted in Fig. 3.3 left and 80.1 J in the right one.

As expected, the expansion is faster in the plasma created with a higher energy laser beam. The hydrodynamic expansion aims at creating the optimal conditions in the plasma to enhance the absorption of the main laser pulse. Figure 3.4 shows vertical cuts (x = 0 μm) of electron density, electron temperature, mean ionization and the absorbed power of the main laser beam ($\lambda = 1.315$ μm, 350 ps duration and 455 J focused in a 130 μm × 4 cm line). These values strongly differ from one case

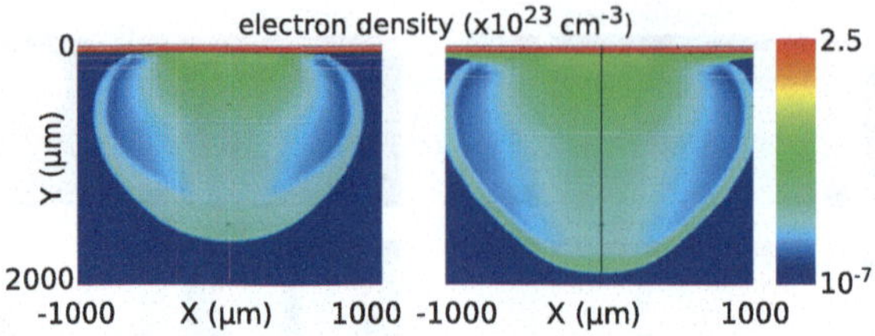

Fig. 3.3 Electron density of Titanium plasma generated by a $\lambda = 1.315\ \mu m$, 300 ps duration pulse focused in a 700 $\mu m \times 4$ cm line. The energy was 32.4 J (left) and 80.1 J (right). The image is taken 7.4 ns after the arrival of the pulse

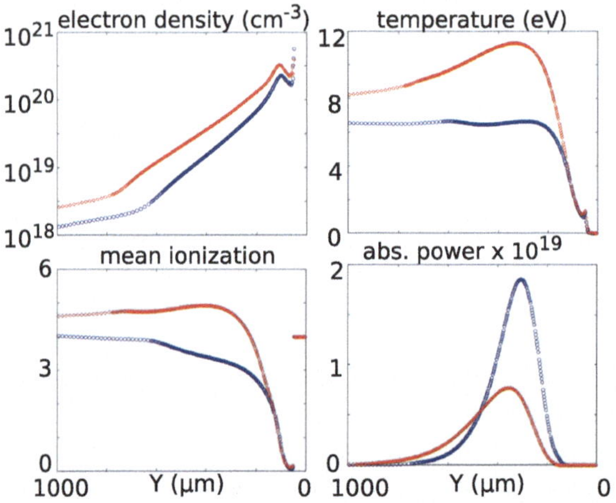

Fig. 3.4 Electron density, electron temperature, mean ionization and absorption of the main laser pulse in a Titanium plasma generated by a $\lambda = 1.315\ \mu m$, 300 ps duration pulse focused in a 700 $\mu m \times 4$ cm line. The energy was 32.4 J (left) and 80.1 J (right). The image is taken 7.4 ns after the arrival of the pulse

to the other (the only difference is the energy of the prepulse) and the absorption profile of the main pulse is also different.

Using a higher energy prepulse (red lines in Fig. 3.4) creates a more homogeneous plasma, at least in the central part, compared with the lower energy prepulse plasma (blue lines in Fig. 3.4). The absorption of the main pulse is consequently more homogeneous in the latter case.

It is worth mentioning that these simulations have been performed using a local thermodynamic equilibrium (LTE) model for the mean ionization. This model is valid for the results presented in this paper, since the plasma evolves for several nanoseconds after the arrival of the laser pulse, attaining LTE. However, for shorter characteristic times, a non-LTE model is required to obtain the correct values of mean ionization and electron density [12]. Indeed, it is the case when, in the simulations presented in this paper, the main laser pulse arrives creating the lasing ion and the population inversion.

3.3 Conclusions

In this paper, we have shown ongoing hydrodynamic simulations of QSS plasma-based amplifiers of soft X-rays. Selected results among the several materials that have been simulated (Molybdenum, Silver, Titanium, Zinc) have been presented. The importance of radiative loses has been assessed, being some materials (Zinc) more optimal for QSS amplifiers than others (Molybdenum). In addition to this, we presented some results of the optimization of the prepulse that creates the QSS amplifier when using Titanium as a target material.

Acknowledgements This work has received funds from the Universidad Politécnica de Madrid *Programa propio* projects VJOVUPM18EOG and VJIDOCUPM18EOG DERKETA and the Czech Science Foundation project GA18-27340S.

References

1. M.J. Berger, P. Colella, Local adaptive mesh refinement for shock hydrodynamics. J. Comp. Phys. **82**, 64–84 (1989)
2. K. Cassou, P. Zeitoun, P. Velarde, F. Roy, F. Ogando, M. Fajardo, G. Faivre, D. Ros, Transverse spatial improvement of a transiently pumped soft-x-ray amplifier. Phys. Rev. A **74**, 045802 (2006)
3. M. Cotelo, P. Velarde, A. de la Varga, D. Portillo, C. Stehl, U. Chaulagain, M. Kozlova, J. Larour, F. Suzuki-Vidal, Simulation of radiative shock waves in xe of last pals experiments, in *10th International Conference on High Energy Density Laboratory Astrophysics*. High Energy Density Physics, vol. 17 (2015), pp. 68–73
4. M. Cotelo, P. Velarde, A.G. de la Varga, C. García-Fernández, Equation of state for laboratory astrophysics applications. Astrophys. Space Sci. **336**(1), 53–59 (2011)
5. A. de la Varga, P. Velarde, M. Cotelo, F. de Gaufridy, P. Zeitoun, Radiative properties for warm and hot dense matter. High Energy Density Phys. **7**(3), 163–168 (2011)
6. A. Depresseux, E. Oliva, J. Gautier, F. Tissandier, J. Nejdl, M. Kozlova, G. Maynard, J.-P. Goddet, A. Tafzi, A. Lifschitz, H.T. Kim, S. Jacquemot, V. Malka, K. Ta Phuoc, C. Thaury, P. Rousseau, G. Iaquaniello, T. Lefrou, A. Flacco, B. Vodungbo, G. Lambert, A. Rousse, P. Zeitoun, S. Sebban, Table-top femtosecond soft x-ray laser by collisional ionization gating. Nat. Photonics. 9(12), 817 (2015)
7. C.G. Durfee, H.M. Milchberg, Light pipe for high intensity laser pulses. Phys. Rev. Lett. **71**, 2409–2412 (1993)
8. D.C. Eder, P. Amendt, S.C. Wilks, Optical-field-ionized plasma x-ray lasers. Phys. Rev. A **45**, 6761–6772 (1992)
9. D. García-Senz, P. Velarde, F. Suzuki-Vidal, C. Stehlé, M. Cotelo, D. Portillo, T. Plewa, A. Pak, Interaction of hemispherical blast waves with inhomogeneous spheres: probing the collision of a supernova ejecta with a nearby companion star in the laboratory. Astrophys. J. **871**(2), 177 (2019)
10. P.V. Nickles, V.N. Shlyaptsev, M. Kalachnikov, M. Schnurer, I. Will, W. Sandner, Short pulse x-ray at 32.6 nm based on transient gain in ne-like titanium. Phys. Rev. Lett. **78**, 2748 (1997)
11. F. Ogando, P. Velarde, Development of a radiation transport fluid dynamic code under amr scheme. J. Quant. Spectrosc. Radiat. Transf. **71**, 541 (2001)
12. E. Oliva, A. Depresseux, M. Cotelo, A. Lifschitz, F. Tissandier, J. Gautier, G. Maynard, P. Velarde, S. Sebban, Hydrodynamic evolution of plasma waveguides for soft-x-ray amplifiers. Phys. Rev. E **97**, 023203 (2018)

13. E. Oliva, P. Zeitoun, P. Velarde, M. Fajardo, K. Cassou, D. Ros, S. Sebban, D. Portillo, S. le Pape, Hydrodynamic study of plasma amplifiers for soft-x-ray lasers: a transition in hydrodynamic behavior for plasma columns with widths ranging from 20 μm to 2 mm. Phys. Rev. E **82**, 056408 (2010)
14. G.J. Pert, The hybrid model and its application for studying free expansion. J. Fluid Mech. **131**(401) (1983)
15. B. Rus, T. Mocek, A.R. Präg, M. Kozlová, G. Jamelot, A. Carillon, D. Ros, D. Joyeux, D. Phalippou, Multimilijoule, highly coherent x-ray laser at 21 nm operating in deep saturation through double-pass amplification. Phys. Rev. A **66**(6), 063806 (2002)
16. P. Velarde, D. García-Senz, E. Bravo, F. Ogando, A. Relaño, C. García, E, Oliva, Interaction of supernova remnants: from the circumstellar medium to the terrestrial laboratory. Phys. Plasmas **13**(9), 092901 (2006)
17. P. Velarde, F. Ogando, S. Eliezer, J. Mart-nez-Val, J. Perlado, M. Murakami, Comparison between jet collision and shell impact concepts for fast ignition. Laser Part. Beams **23**(1), 43–46 (2005)

Chapter 4
The Use of Inhomogeneous Forming Lines in the Laser Power Supply System on a High-Current Z-Discharge Plasma

N. Kalinin, V. Antonov, S. Vaganov and A. Kovalenko

Abstract The paper presents the numerical simulation results of the energy characteristics of a laser based on no equilibrium plasma of multiply charged ions. Plasma is created in a small inductive extended Z-discharge with a power system on heterogeneous forming lines. The requirements for the distribution of the lines parameters providing the creation of the energy density and power density necessary for the generation of radiation in the "water window" region have been established.

4.1 Introduction

An important problem that limits the advancement of high-spectral brightness sources to the extreme ultraviolet—soft X-ray (EUV-MR) spectral range on the plasma of multiply charged ions in the region of shorter wavelengths is the creation of active medium pump sources that provide the necessary high specific power input into the plasma. To obtain the generation of radiation in the range of 1–10 nm, the specific pump powers of at least 10 TW/cm^3 are necessary. In solving this problem, out of competition are powerful pulse laser systems. At the same time, there is still interest in the creation and study of EUV-MR emission sources with the pumping of the active medium using high-current capillary-type electrical Z-discharges [1–4].

In this paper, an analysis is made of how to create an effective energy input into the plasma of high-current capillary Z-discharges. Plasma sources of radiation pumped by high-current Z-discharges have the following merit.

The efficiency of energy input into the plasma from the primary energy storage and the amount of energy transferred to the plasma can be greater than in laser systems. To do this, it is necessary to match the rise time of the discharge current and the characteristic plasma compression time. An important mechanism for transferring

N. Kalinin · S. Vaganov · A. Kovalenko
Joffe Physical-Technical Institution Russian Academy of Science, St. Petersburg, Russia

V. Antonov (✉)
Peter the Great St. Petersburg Polytechnic University, St. Petersburg, Russia
e-mail: antonovvi@mail.ru

M. Kozlová and J. Nejdl (eds.), *X-Ray Lasers 2018*, Springer Proceedings in Physics 241, https://doi.org/10.1007/978-3-030-35453-4_4

energy to the plasma is the work of the magnetic field created by the flowing current [5–9].

The compression of the plasma can begin even with a small discharge current. When the discharge current reaches high values, the force accelerating the plasma shell decreases due to the decrease in the plasma radius. The introduction of energy into the plasma can be of little effect if the plasma shell converges on the axis of the discharge chamber before the current reaches its maximum value. In this case, it is necessary to reduce the rise time of the discharge current by the two ways considered in our work.

1. By using forming lines with homogeneous and inhomogeneous distributions of characteristics,
2. By profiling the initial plasma density distribution in the discharge chamber.

4.2 Mathematical Model

The efficiency of the energy input into the plasma with a high-current discharge can be estimated using the "snow plow" model. In this case, it is necessary that the condition $R_{\mathrm{pl}}/R_0 > 0.2 - 0.3$ be fulfilled.

The magnetic field of the discharge current does not penetrate deep into the plasma. The electrical conductivity of the plasma is infinitely large so that the radial distribution of the magnetic field pressure B_φ has the character of an ideal piston. The plasma layer moves to the axis of the discharge chamber with a magnetic piston in the form of a thin layer of variable mass in the form of a "snow plow". The mass of the plasma layer is

$$M_{pl} = \pi\rho\left(R_0^2 - R_{pl}^2\right) \tag{4.1}$$

where ρ is the initial plasma density, R_{pl} is the radius of an infinitely thin plasma shell. The equations of the plasma shell motion have the following form:

$$\begin{gathered} 2\pi R_{pl} P_B = -\tfrac{d}{dt}\left(M_{pl}\tfrac{dR_{pl}}{dt}\right); \\ \tfrac{d}{dt}\left[\rho\left(R_0^2 - R_{pl}^2\right)\tfrac{dR_{pl}}{dt}\right] = -\tfrac{\mu_0}{4\pi}\tfrac{I(t)^2}{R_{pl}} + 2\pi R_{pl} P; \end{gathered} \tag{4.2}$$

$$B_\phi = \frac{\mu_0}{2\pi}\frac{I(t)}{R_{pl}(t)}; \quad P_B = \frac{B_\phi^2}{2\mu_0}.$$

Here $I\ (t)$ is the discharge current, B_φ is the magnetic field, P_B is the pressure of the magnetic piston.

After a while, the initial gas-kinetic pressure in the discharge chamber will begin to affect the plasma compression process. It must be taken into account in the equation of motion of the plasma shell. To do this, a term appears on the right-hand side of the equation that relates the change in the gas-kinetic pressure with the change in the radius of the plasma shell. If we assume that the gas-kinetic pressure is uniformly distributed along the radius of the plasma column, then this corresponds to adiabatic compression of the ionized gas $PV^{\ddot{\imath}} = \text{const}$. For a contracting cylinder $V \sim R_{\text{pl}}^2$, therefore $P = P_0(R_{\text{pl}}/R_0)^{-2\gamma}$.

The energy introduced into the plasma is determined by the work done by the magnetic piston in plasma compression

$$W = \pi \int_{R_0}^{R_{pl}} \frac{B_{\varphi}^2}{\mu_0} R_{\text{pl}} dr \tag{4.3}$$

A voltage pulse generator through an inhomogeneous line is the source of a little inductive capillary-type Z-discharge. The design of the line makes it possible to form voltages or currents of the required shape at an arbitrary load.

We have considered the power supply of the capillary-type Z-discharge from the generator of voltage pulses through a heterogeneous line. For comparison, the case of a discharge plasma load directly from the condenser battery and the case of plasma load connection to the voltage generator through a homogeneous line is considered (Fig. 4.1).

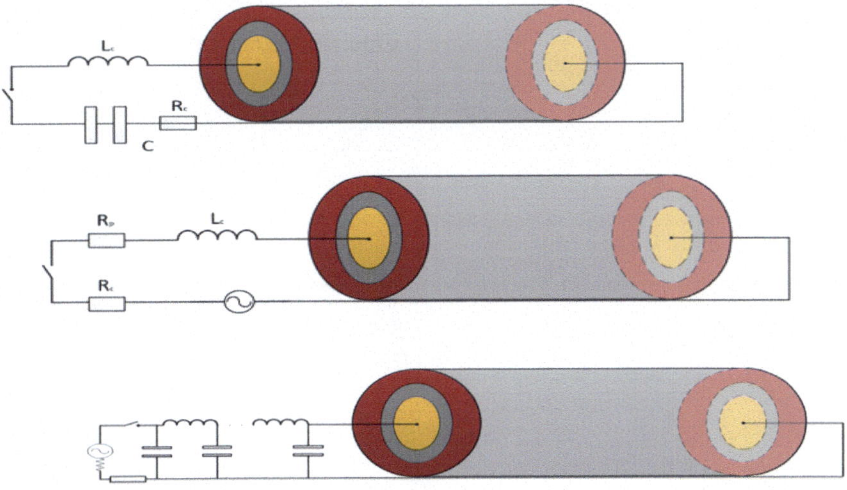

Fig. 4.1 Discharge circuits

Differential equations of an inhomogeneous line without losses on the conductivity

$$-\frac{\partial U}{\partial y} = W(y)\frac{\partial I}{\partial t}; \quad -\frac{\partial I}{\partial y} = \frac{1}{W(y)}\frac{\partial U}{\partial t}$$
$$V(x) = \frac{1}{\sqrt{L(x)C(x)}}; \quad y(x) = \int_0^x \frac{d\xi}{V(\xi)}; \quad W(y) = \sqrt{\frac{L(x)}{C(x)}} \tag{4.4}$$

Here $L(x)$, $C(x)$ are the given inductance and capacitance distributions.

$\frac{\partial I}{\partial t} = \frac{U_g(t)}{L_0}$ on the left border of the line, y = 0;

$U = I\frac{dL_{pl}}{dt} + L_{pl}\frac{dI}{dt}; \quad L_{pl}(R_{pl}) = \frac{\mu_0}{2\pi}\mathrm{Ln}\left(\frac{R_0}{R_{pl}}\right)$ on the right border.

The problem was solved numerically by the method of "straight lines".

The results of the calculations are presented in Figs. 4.2, 4.3 and 4.4 in relative values. The numbers in these figures mean the following.

Figure 4.1. 1. Voltage on condenser battery; 2. Discharge current; 3. Plasma shell radius; 4. Plasma shell velocity.

Figures 4.2 and 4.3. 1. Discharge current; 2. Plasma shell radius; 3. Plasma shell velocity.

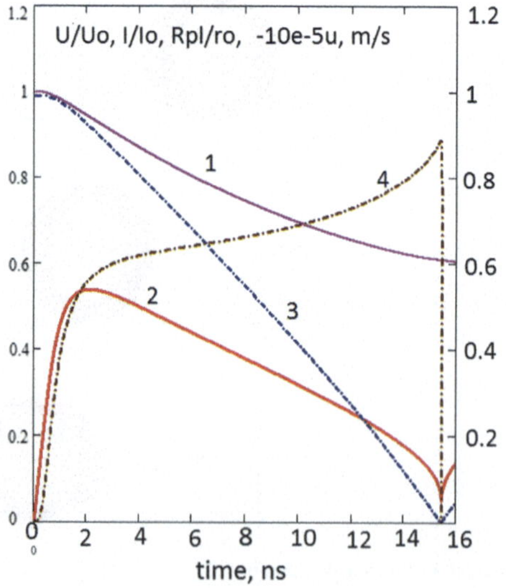

Fig. 4.2 Discharge characteristics. First scheme

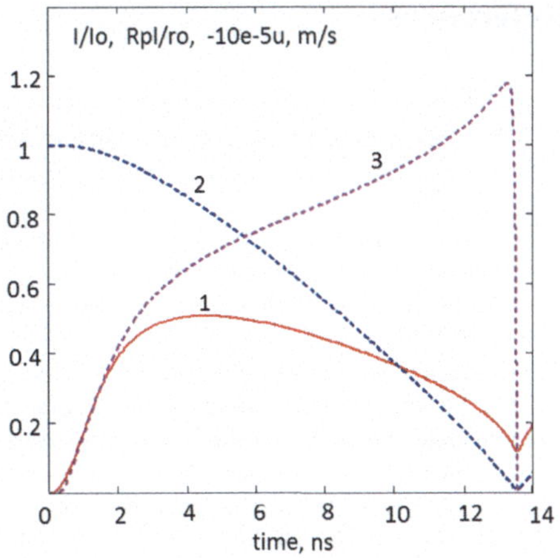

Fig. 4.3 Discharge characteristics. Second scheme

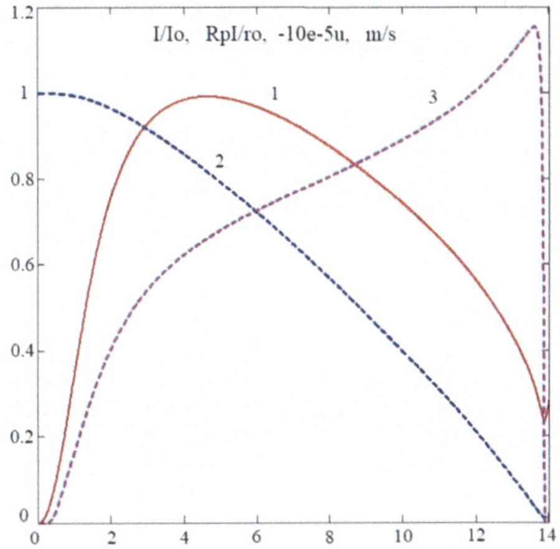

Fig. 4.4 Discharge characteristics. Third scheme

4.3 Conclusions

The efficiency of plasma heating in high-current Z-discharges I is largely determined by the possibility of the current pulse formation with a high slew rate in the plasma load. Power supply systems for high-current Z-discharges based on high-voltage generators and forming lines are considered. They allow the creation of current pulses in a discharge with a higher slew rate than in the case of capacitor batteries. In this case, the maximum plasma compression rate at the time of implosion on the axis and the efficiency of energy input into the plasma is greater than when the discharge is powered by a capacitor battery. The use of lines with a nonuniform distribution of wave resistance enhances this effect. So, for example, a line with a linear drop in wave resistance makes it possible to increase the maximum plasma implosion speed by about 30% compared to a source in the form of a battery.

References

1. R.C. Elton, *X-ray Lasers* (Academic Press)
2. A.V. Vinogradova, J.J. Roccab, P.N. Lebedev, Repetitively pulsed X-ray laser operating on the 3p–3s transition of the Ne-like argon in a capillary discharge. Quantum Electron. **33**(1), 7–17 (2003)
3. A. Jancarek, L. Pina, M. Vrbova et al., Nitrogen capillary discharge emission in 1.9–2.5 nm wavelength range. Czechoslovak J. Phys. **56**, 250–254 (2006)
4. A. Kolb, G. Grim, High-temperature shock waves Achievements of physical sciences. **82**(1), 83–131 (1964)
5. I.G. Kataev, Shock electromagnetic waves (Soviet Radio, Moscow, 1963)
6. O.N. Litvinenko, V.I. Soshnikov, The theory of inhomogeneous lines and their application in radio engineering (Soviet Radio. Moscow, 1964)
7. V.A. Burtsev, Y.L. Ermolaev, N.V. Kalinin, Future possible application of different powerful pulsed generators in accelerating technology, in *Proceedings of the 8th International Conference on High-Power Particle Beams,* vol. 2 (BEAMS-90, Novosibirsk, USSR, 1990), pp. 996–1009
8. H. Ghomi, M. Yousfli, S.H. Rostami, Y. Hayashi, E. Hotta, Optimum driving a Z-pinch for X-ray lasers. Int. J. Opt. Photonics (IJOP) **4**(2), 87–93 (2010)
9. Y.P. Zhao, S.A. Jiang, Y. Xie, Q. Wang, Influence of Z-pinch evolution on laser pulse duration at 46.9 nm in Ne-like ions. Appl. Phys. B. Lasers Opt. **99**, 535–542 (2010)

Chapter 5
Enhanced Fluorescence from X-Ray Line Coincidence Pumping

J. Nilsen, D. Burridge, L. M. R. Hobbs, D. Hoarty, P. Beiersdorfer, G. V. Brown, N. Hell, D. Panchenko, M. F. Gu, A. M. Saunders, H. A. Scott, P. Hatfield, M. P. Hill, L. Wilson, R. Charles, C. R. D. Brown and S. Rose

Abstract Many resonant photo-pumped X-ray laser schemes that use a strong pump line such as Ly-α or He-α to populate the upper laser state of a separate lasing material have been proposed over the last four decades but none have been demonstrated. As a first step to creating a photo-pumped X-ray laser we have decided to reinvestigate some of these schemes at the Orion laser facility with the goal to show enhanced fluorescence. In particular we look at using the Ly-α or He-α K lines to pump the 1s–3p and 4p transitions in H-like Cl and see fluorescence on the 4f–3d line at 65 Å and the 3d–2p line at 23 Å. Preliminary experiments are presented that show a modest enhancement. As an alternative we also look at enhancing the 2p–2s line in Ne-like Ge at 65 Å using the Ly-α Mg line to photo-pump the 2s–3p line of Ne-like Ge. Calculations are presented that suggest modest enhancements of 2.5.

5.1 Introduction

One early approach for creating X-ray lasers was the idea of a resonantly photo-pumped laser where a strong emission line in one material could be used to photo-excite a transition in another material and create lasing. The Na-pumped Ne X-ray laser scheme proposed 40 years ago by Vinogradov and colleagues [1, 2] is a classic example that used the strong Na He-α line at 1127 eV to resonantly photo-pump the Ne He-γ line and lase on the 4f–3d transition at 231 Å in He-like Ne. This scheme was studied extensively and weak gain [3] was inferred in several experiments. There

J. Nilsen (✉) · P. Beiersdorfer · G. V. Brown · N. Hell · D. Panchenko · A. M. Saunders · H. A. Scott
Lawrence Livermore National Laboratory, Livermore, CA 94551, USA
e-mail: nilsen1@llnl.gov

D. Burridge · L. M. R. Hobbs · D. Hoarty · M. P. Hill · L. Wilson · R. Charles · C. R. D. Brown
AWE Plc, Aldermaston, UK

M. F. Gu
University of California, Berkeley, CA, USA

P. Hatfield · S. Rose
Imperial College, London, UK

M. Kozlová and J. Nejdl (eds.), *X-Ray Lasers 2018*, Springer Proceedings in Physics 241, https://doi.org/10.1007/978-3-030-35453-4_5

are a series of similar resonances [2] that use Ly-α or He-α lines to photo-pump Ly-γ or He-γ lines and lase on the 4f–3d lines of H or He-like ions. In this paper we discuss the next higher Z pair of possible resonances that use the strong Ly-α line at 3699 eV from H-like K to resonantly photo-pump the 1s–4p transition in Cl and enhance the emission of the 4f–3d line at 65 Å in H-like Cl.

One can also pump the Ly-β or He-β lines and lase or enhance the fluorescence of the 3d–2p lines. The K–Cl combination is interesting because it also has a resonance between the strong K He-α line to resonantly photo-pump the Cl Ly-β transition and enhance the emission of the 3d–2p line at 23 Å in H-like Cl.

Experiments were done at the Orion laser facility in the UK to look at enhanced emission on both the 23 and 65 Å Cl lines.

An alternative experiment is being designed to look at the emission of the 2p–2s line in Ne-like Ge at 65 Å that could be photo-pumped by the Ly-α line of H-like Mg at 1472 eV. In this case the Ly-α line photo-pumps the inner shell 2s–3p ground state transition. This resonance was measured in previous experiments at the EBIT facility at LLNL, the NPO laser facility at VNIIFTRI, and the Princeton Large Torus (PLT) tokamak. Experiments are being planned at the Orion laser facility to investigate this scheme.

5.2 K-Pumped Cl Experiments

Figure 5.1 shows the energy level diagram for the Potassium (K)-pumped Chlorine (Cl) schemes. The 4f–3d line at 64.8 Å in H-like Cl is enhanced by having the strong Ly-α line of K photo-pump the 1s–4p (Ly-γ) transition of H-like Cl. The n = 4 states

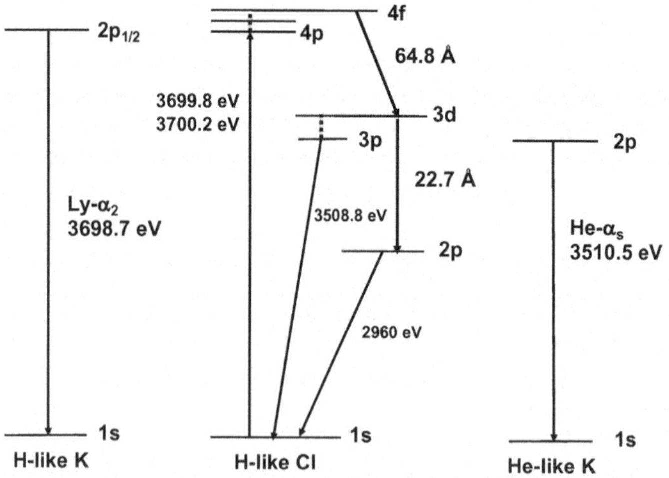

Fig. 5.1 Energy level diagram for the Ly-α and He-α lines of K photo-pumping the 1s–3p and 4p lines of H-like Cl to enhance the fluorescence of the 4f–3d line at 64.8 Å and the 3d–2p line at 22.7 Å

of Cl are all very close in energy and become statistically populated by collisional processes at the temperatures and densities expected in the plasma. The 4f level has the largest statistical weight (2J +1) which results in strong emission and potential gain on the 4f–3d line. Potassium is rather unique because the He-α line, which is usually the strongest line one can produce in a hot plasma, is nearly resonant with the 1s–3p (Ly-β) line of Cl. This photo-pumps population into the 3p level which then equilibrates through collisional processes to statistically populate the n = 3 level resulting in a large population in the 3d level. One then looks for strong emission on the 3d–2p line at 22. 7 Å. This line is also part of the cascade from the 4f–3d emission which also enhances the line. This scheme was originally proposed and modelled [3, 4] many years ago.

To look for potential enhancements, experiments were done at the Orion laser facility [5] at the atomic weapons establishment (AWE) in the UK. The laser systems consist of ten beams, optimized around 1 ns pulse duration, which each provide a nominal 500 J at a wavelength of 351 nm. There are also two short pulse beams, which each provide 500 J in 0.5 ps at 1054 nm, that were not used in these experiments.

The targets consisted a 0.5 μm thick microdot of KCl with a diameter of 300 μm. This was coated on each side with 0.5 μm thick layer of KF and then an outer protective coating of 0.3 μm thick layer of parylene-N, which is a CH plastic polymer. The goal of the KF layer was to have a thick layer of K surrounding the Cl plasma that would enhance the strength of the Ly-α and He-α K lines pumping the Cl. As a null experiment we used targets that replaced the KCl with NaCl and the KF with NaF. In the null experiment there are no K lines to photo-pump the Cl plasma.

The targets were hit from both side by 250–1150 J of energy in a 300 μm diameter beam with a 1–1.5 ns FWHM pulse duration. Hydrodynamic calculations done with the 1D NYM code predict that the plasma would expand and blow down to densities of about 1–2 mg/cc or electron density of 3–6 × 10^{20}/cc and reach temperatures up to several keV. Under these conditions we expect to ionize the plasmas to H and He-like states. We observe strong emission, as seen in Fig. 5.2, on the K and Cl Ly-α and He- α lines for the nominal target and strong emission on the Cl Ly-α and He-α lines for the null target indicating that we have reached the correct ionization conditions. This data was recorded on the Mark 2 time-integrated spectrometer with a resolution of 7 eV using CsAP crystals.

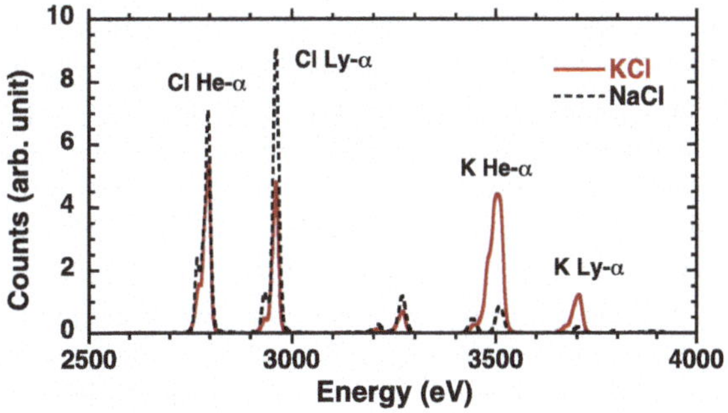

Fig. 5.2 Emission from the NaCl target and KCl target showing the strong Ly-α and He-α emission lines from K and Cl

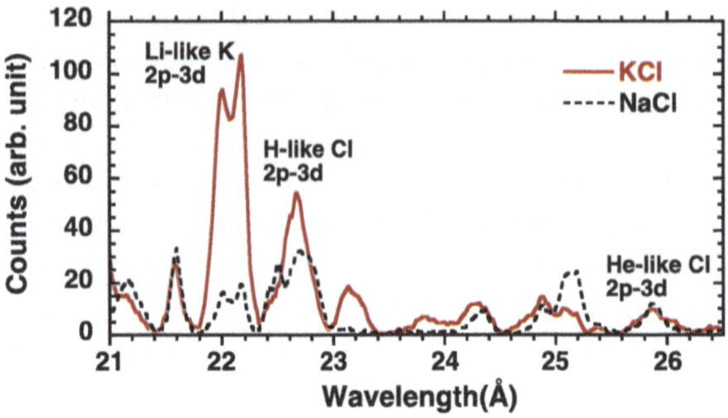

Fig. 5.3 Emission from the NaCl and KCl targets shows modest enhancement of the H-like Cl 3d–2p line at 22.7 Å in the KCl target

To look at the spectral lines near 23 and 65 Å we used a grating instrument spectrometer that covered the range of 100–1000 eV (120–12 Å) with a resolution of 1000. Figure 5.3 shows the spectrum from 21–27 Å that looks at the enhancement of the Cl 3d-2p line at 22.7 Å. When we normalize to the other Cl lines we believe we observe a modest enhancement of about 20% on this line due to the photo-pumping. The spectrum near the 65 Å line is more complicated and still being analysed as there are nearby lines in Li-like Cl in the same spectral region.

5.3 Mg-Pumped Ge Experiments

The second scheme that we are actively pursuing, shown in Fig. 5.4, uses the Magnesium (Mg) Ly-α lines to photo-pump the Neon (Ne)-like Germanium (Ge) 2s–3p line (B) and enhance the emission of the 2p–2s line in Ne-like Ge at 65.1 Å. The resonance between the Mg and Ge lines were measured in previous experiments at the EBIT facility at LLNL, the NPO laser facility at VNIIFTRI, and the Princeton Large Torus (PLT) tokamak [6]. Experiments are being planned at the Orion laser facility to investigate this scheme.

These experiments would be very similar to the KCL experiments where we use microdot targets with a 0.2 μm thick Ge target surrounded by 0.3 μm of Mg for the pumped experiment and 0.3 μm of Al for the null experiment. Both sets of targets are overcoated with 0.3 μm of Parylene-N on each side as a protective layer from the atmosphere. To investigate the viability of this scheme we have done a series of kinetics and radiation transport calculations with the Cretin code [7] to estimate enhancements that we might see in the upcoming experiments at the Orion facility.

If we assume the targets have blown down in density to an electron density of 5 $\times$ 10^{20}/cc (1.8 mg/cc) at a temperature of 500 eV we can first estimate the strength and width of the Mg Ly-α alpha line. Doing a 1D line transport calculation assuming a Mg target with a radius of 350 μm, Fig. 5.5 shows the line strength for the Mg Ly-α in the center of the target. One sees that the line has a width of 2.1 eV with the Ge line located at an energy of 0.6 eV relative to the Ly-α2 line. This makes for essentially a perfect resonance between the Mg and Ge lines. The Mg line has a peak

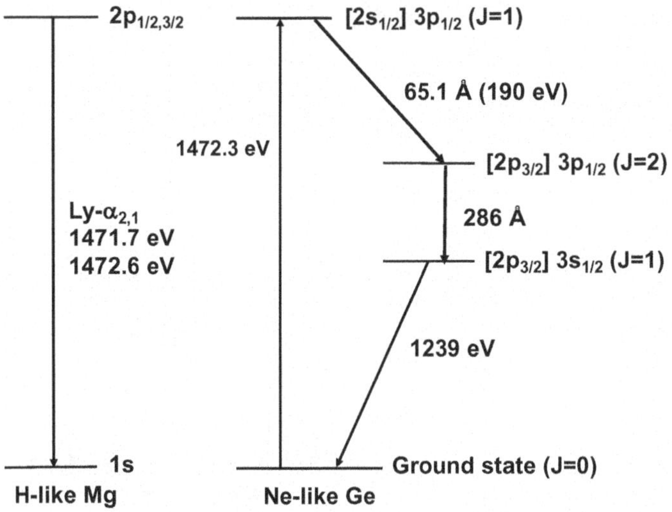

Fig. 5.4 Energy level diagram for the Mg Ly-α photo-pumping the Ne-like Ge 2s–3p line to enhance the fluorescence of the 2p–2s line at 65.1 Å

Fig. 5.5 Line strength for the Mg Ly-α line versus photon energy centered around the Ly-α2 line

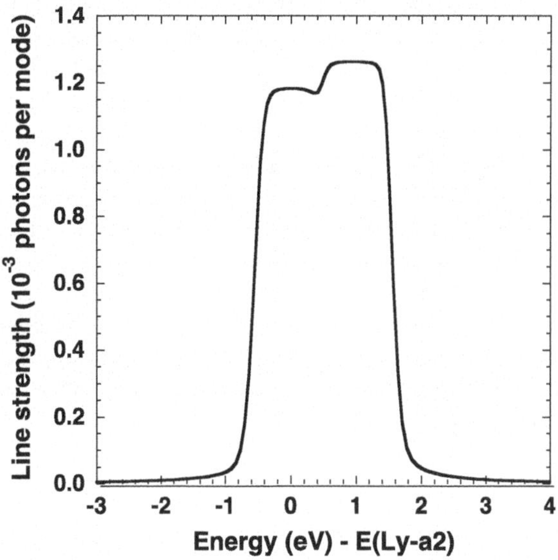

strength of 1.2×10^{-3} photons per mode in the center of the target and falls off by a factor of 6 at the outer surface of the target. When pumping a line, the line strength in photons per mode is very similar to the fractional population of the upper state being pumped.

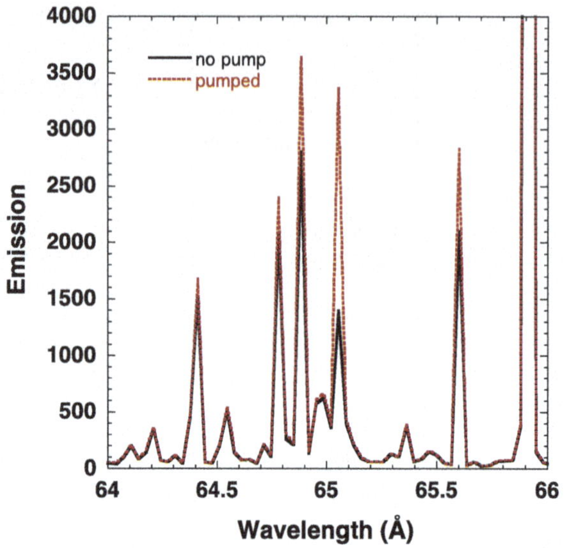

Fig. 5.6 Predicted emission from the Ge target comparing the case where it is pumped by the Mg Ly-α line to the case without any pumping

The next step is to do a kinetics calculation for the spectra of Ge near 65 Å with and without the presence of the Mg line using the pump strength calculated above. To best estimate the spectra, we created an atomic model for Ge that uses many body perturbation theory calculations for the energy levels of Ne-like Ge $n = 2$ and $n = 3$ levels. These calculations have been compared with detail spectra taken at the EBIT facility [8] at LLNL.

Taking the same plasma conditions described above for Ge we use Cretin to calculate the Ge spectra. Without the Mg pump present we observe a series of lines that are dominated by the very strong F-like Ge 2p–2s line at 65.9 Å. When we do a similar calculation including the Mg pump line the spectra are very similar except for a significant enhancement of about 2.5 on the 2p–2s Ne-like Ge line at 65.06 Å, as shown in Fig. 5.6. This is the type of enhancement we are trying to observe in the future experiments. One advantage of looking for enhanced fluorescence instead of gain is that it can take a more substantial enhancement of the line before it has enough gain to lase.

5.4 Conclusions

In this paper we look at using the K Ly-α or He-α lines to photo-pump the 1s–3p and 4p transitions in H-like Cl and see fluorescence on the 4f–3d line at 64.8 Å and the 3d–2p line at 22.7 Å. Preliminary experiments are presented that show a modest enhancement of about 20% on the 22.7 Å line. As an alternative we also look at enhancing the 2p–2s line in Ne-like Ge at 65.1 Å using the Mg Ly-α line to photo-pump the 2s–3p line of Ne-like Ge. Calculations are presented that suggest modest enhancements of 2.5. Experiments are planned in the near future to test the Mg-pumped Ge scheme.

Acknowledgements This work was performed under the auspices of the U.S. Department of Energy by Lawrence Livermore National Laboratory under Contract DE-AC52-07NA27344 in collaboration with the Atomic Weapons Establishment of the United Kingdom.

References

1. A.V. Vinogradov, I.I. Sobelman, E.A. Yukov, Sov. J. Quantum Electron. **5**, 59–63 (1975)
2. J. Nilsen, J.H. Scofield, E.A. Chandler, Appl. Opt. **31**, 4950–4956 (1992)
3. J. Nilsen, E.A. Chandler, Phys. Rev. A **44**, 4591–4598 (1991)
4. S.J. Rose, J. Phys. B **31**, 2129–2144 (1998)
5. N. Hopps et al., Plasma Phys. Control Fusion **57**, 064002 (2015)
6. J. Nilsen et al., Phys. Rev. A **50**, 2143 (1994)
7. H.A. Scott, JQSRT **71**, 689–701 (2001)
8. P. Beiersdorfer, Nuclear Instrum. Methods Phys. Res. B **205**, 173 (2003)

Chapter 6
Generation of Short Wavelength Radiation and Energetic Particles by Using PW-Class Lasers

T. Kawachi, M. Nishikino, A. Sasaki, N. Hasegawa, M. Ishino, T.-H. Dinh, K. Mikami, T. Kitamura, H. Kiriyama, A. S. Pirozhkov, T. Th. Esirkepov, A. Sagisaka, M. Nishiuchi, H. Sakaki, M. Kando, A. Ya. Faenov, T. A. Pikuz, K. Kawase, R. Nagai, R. Hajima, S. V. Bulanov and K. Kondo

Abstract This paper describes recent progress of the study of short wavelength coherent radiation and laser particle acceleration by the use of ultra-intense lasers in our institute. In laser-driven soft x-ray laser research, the 18.8 nm laser has been demonstrated with the repetition rate of 10 Hz, and new driver laser with higher repetition rate is under development. Generation of higher order harmonics from relativistic plasmas is conducted by using J-KAREN-P, and the source size together with the spatial coherence are investigated. We also show the progress in laser-driven heavy ion acceleration study, which is done as a part of the development of compact cancer therapy machine, and the proposal of alternative method toward high average power coherent attosecond x-ray generation.

T. Kawachi (✉) · M. Nishikino · A. Sasaki · N. Hasegawa · M. Ishino · T.-H. Dinh · K. Mikami · T. Kitamura · H. Kiriyama · A. S. Pirozhkov · T. Th. Esirkepov · A. Sagisaka · M. Nishiuchi · H. Sakaki · M. Kando · S. V. Bulanov · K. Kondo
Kansai Photon Science Institute (KPSI), National Institutes for Quantum and Radiological Science and Technology (QST), Kyoto, Japan
e-mail: kawachi.tetsuya@qst.go.jp

K. Kawase · R. Nagai · R. Hajima
Takasaki Advanced Radiation Research Institute, National Institutes for Quantum and Radiological Science and Technology (QST), Ibaraki, Japan

A. Ya. Faenov · T. A. Pikuz
Institute for Open and Transdisciplinary Research Initiatives, Osaka University, Osaka, Japan

Joint Institute for High Temperature of the Russian Academy of Science, Moscow, Russia

S. V. Bulanov
ERT/High Field Initiative (HIFI), ELI-Beamlines, Dolní Břežany, Czechia

M. Kozlová and J. Nejdl (eds.), *X-Ray Lasers 2018*, Springer Proceedings in Physics 241, https://doi.org/10.1007/978-3-030-35453-4_6

6.1 Introduction

Kansai Photon Science Institute (KPSI) had been a laboratory on advanced photon science using high peak power lasers under the umbrella of Japan Atomic Energy Agency (JAEA). In 2016, by the change of policy of national measures, KPSI was transferred to National Institutes for Quantum and Radiological Science and Technology (QST), which was newly established to promote quantum science and its applications to material science, biological science, and fusion science. In QST, the main subjects of KPSI do not change and are, based upon ultra-intense laser technologies, the laser-driven secondary sources toward medical and industrial applications and ultrafast probe for material and biological substances [1].

The study of coherent x-rays such as plasma-based soft x-ray lasers (SXRLs) and higher order harmonics generation (HHG) from neutral atoms or plasma electrons is very important not only as an interesting objectives of high field science but also as a material and biological probe. Therefore we are putting great efforts into this research field.

In this paper, we introduce the recent progress and future prospect in the source development of SXRLs in transient collisional excitation (TCE) scheme with high repetition-rate operation and shorter wavelength HHG from plasma by use of our PW-class laser, J-KAREN-P (Petawatt). In addition to this, we also describe other research activities in KPSI: One is the study of laser-driven heavy ion acceleration for hadron therapy, and another is a proposal to generate very high average power attosecond x-rays by use of a combination of conventional electron accelerator and carrier-envelope phase (CEP) stabilized laser technology.

6.2 Research on TCE Lasers

After the demonstration of a fully spatial coherent x-ray laser with TCE scheme, we developed TOPAZ laser as the SXRL driver laser [2]. TOPAZ employed a unique zigzag slab amplifier operated with 0.1 Hz repetition rate. By use of SXRLs pumped by TOPAZ, we have conducted several application experiments to pursue material science and EUV laser processing. However, the repetition rate of 0.1 Hz became insufficient when we advanced the applications such as pump-probe measurement, therefore we decided to replace TOPAZ with a way that higher repetition-rate, e.g., 10 or 100 Hz, can be expected.

As a start, we have confirmed the lasing at 18.8 nm in the nickel-like Mo ions with the grazing incidence pump (GRIP) scheme (See Fig. 6.1a). The current driver is a Ti:Sapphire laser with the pump energy of 700 mJ on the target under 10 Hz operation. We plan to modify the current laser system into more powerful and versatile and easily accessible system. In the pulse energy, an additional pump laser will be added and the output energy will increase up to 2 J on target, which will make it possible to generate fully spatial coherent 13.9 nm laser by coupling with HHG seeder. For

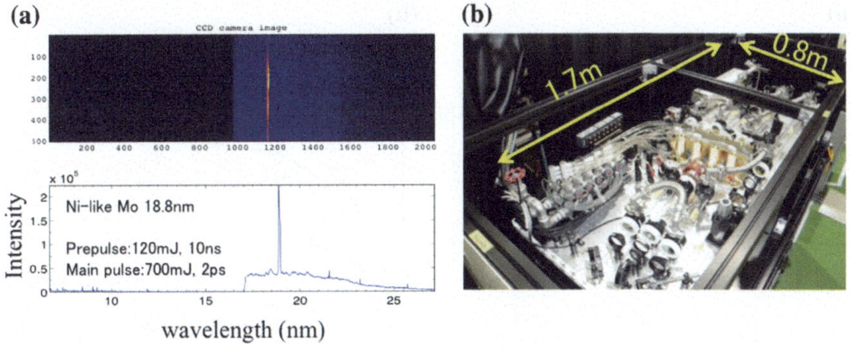

Fig. 6.1 Typical amplification line from Nickel-like Mo ions pumped by 10 Hz repetition-rate Ti:Sapphire laser; (**a**). Internally developed Nd:YAG laser system which can provide laser pulses with the energy of 4.7 J under 50 Hz operation; (**b**)

higher repetition rate, we are also developing a Nd:YAG pump laser, which can work with a repetition-rate up to 100 Hz. The laser-diode pump oscillator provides 15 ns duration seed pulse, and the seed is amplified by an internally developed flash-lamp-pumped amplifier incorporating 3 Nd:YAG rods arranged to be able to compensate for the nonuniformity of the flash-lamp illumination. By adjusting and optimizing the image relay, thermal lens effect is avoided as much as possible, and we can obtain the output laser pulse with 4.7 J (fundamental) which is quite stable under 50 Hz operation [3]. Our evaluation of the thermal distortion effects indicates that 100 Hz operation is possible by this system. This new pumping laser is very compact and robust: the unit size is only 1.7 m × 0.8 m (Fig. 6.1b), and our next plan is to test the second harmonics generation with this laser.

6.3 Higher Order Harmonics from Relativistic Plasma Using PW-Laser

In the interaction of ultra-intense laser with gas target, plasma electrons originated from atoms are swept out by the ponderomotive force, and the electron flow forms a Wakefield. Under a laser irradiation condition where self-focusing occurs in plasma, the electron flow starts to be detached from the ion cavity and becomes bow wave. At this detached point, the electron flow stagnates and forms a localized density peak or density singularity. Since this localized area is inside of the laser electric field, the electrons in this area are strongly oscillated collectively, and very strong radiation emits from this point. We call this process burst intensification of singularity emitting radiation (BISER). In BISER, if the size of the emitting area is comparable to the spatial scale of the periodic driving field, the emission, not only the fundamental but also higher order harmonics become spatially and temporally coherent, and this condition will be realized more easily as the laser intensity increases [4].

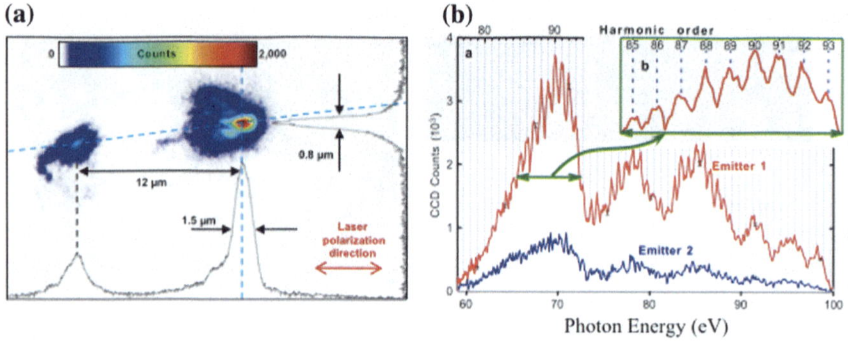

Fig. 6.2 Single-shot image of the source in plasma. Two emitters are corresponding to the localized density peaks in the left- and right-hand side in terms of linearly polarized laser electric field; (**a**). Typical spectra in soft X-ray region taken by the grazing incidence spectrograph; (**b**)

We conducted an experiment using J-KAREN-P laser [5] in KPSI to identify the size of the singularity. The linearly polarized laser with the intensity of 6.0×10^{19} ~ 2.0×10^{20} W/cm^2 irradiated the helium gas target with the electron density of 1.4 ~ 6.0×10^{19} cm^{-3}. We employed X-ray imaging system, which consisted of the normal incidence concave multilayer mirror and LiF detector and a grazing incidence spectrograph. Figure 6.2a shows the single shot image of the source corresponding to the photon energy from 60 eV through 100 eV. We observed two emitters separated by 15 μm., which was almost the same with the laser spot size, ~10 μm. The size of the brighter emitter was around 1.5–0.8 μm including small influence of the aberration, but the size was consistent with the expected spatial scale of the laser electric field.

Figure 6.2b shows the typical spectra from two emitters measured by the grazing incidence spectrograph. We could see clearly sawtooth like fine-fringes and modulation with the period of 5–10 eV. The modulation, although we need further analysis, may be due to interference between two emitter sources or autocorrelation in the spectrograph throughput. For the fine-fringes, the wavelengths completely agree with those of higher order harmonics, and this is the proof that the source is coherent. The total energy of the HHG ranging from 60 to 100 eV was estimated to be 100 nJ/shot, which is corresponding to the photon number of 10^{10}. These values are almost consistent with the theoretical estimation by our 3D-PIC code.

It should be noted that the measured spectral-width indicated the duration of each HHG signal of around 170 as consistent with the theoretical prediction ~150 as. Since the maximum photon energy of BISER is scaled with the laser intensity as $h\omega_L n_{max}$ ~a_0^3, where $h\omega_L$ is the photon energy of the driver laser and a_0 is normalized vector potential, it is expected by using BISER that the photon energy range of coherent attosecond pulses can be extended to several keV range which has not been realized yet with existing X-ray sources.

6.4 Other Research Activities Using J-KAREN-P and Alternative Scheme for Ultrashort X-ray Generation

In this chapter, we describe the recent progress in laser-driven heavy ion acceleration research as another application of J-KAREN-P and a proposal of alternative scheme for the attosecond X-ray generation in which an electron accelerator and CEP locked laser are combined.

Ion beam cancer therapy has a great advantage in concentration of the irradiation effect due to the Bragg peak as compared to X-ray irradiation, and in terms of the linear energy transfer (LET) value in dosimetry, the carbon ion beam is more effective in comparison with a proton beam. Therefore heavy ion cancer therapy is attracting attention as a highly effective method in treating cancer. A branch of QST, National Institute for Radiological Science (NIRS) owns the first-generation heavy ion cancer therapy machine HIMAC, and conducts approximately 1000 cancer treatments annually. However, since the size of the device is large, the treatment cost is inevitably high in terms of the maintenance cost of the device, and downsizing of the device is an urgent issue. Laser ion acceleration is capable to produce an acceleration electric field larger by approximately 5 orders of magnitude than that of the current RF acceleration cavity and has the potential to greatly miniaturize existing heavy-ion accelerators.

Our plan in development of laser ion accelerator is step by step. The tentative goal is to develop a laser-driven ion injector which is coupled with main synchrotron accelerator and gantry using super-conducting magnet. We call this system as Quantum Scalpel. The expected size of this device is 20 m $\times$ 10 m, much smaller than that of the conventional heavy ion accelerator. The required specification for the ion injector is 10^9 of C^{6+} ions in the energy range of 4 MeV/u $\pm$ 10% per mSr in 2 s.

Figure 6.3 shows the typical Thomson parabola spectra of accelerated ions taken in the J-KAREN-P experiment. An SUS foil target with thickness of 5 μm is irradiated

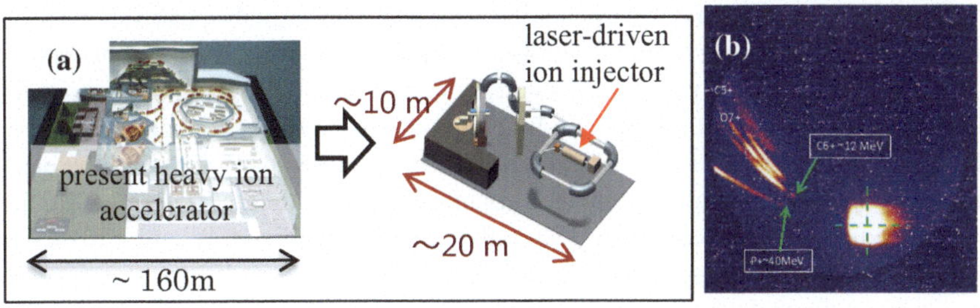

Fig. 6.3 Laser-driven ion injector coupled with synchrotron accelerator using super-conducting magnet will make it possible to realize a compact heavy ion therapy machine "Quantum Scalpel;" (**a**). Typical energetic ion spectra taken by Thomson parabola detector in the experiment using J-KAREN-P; (**b**)

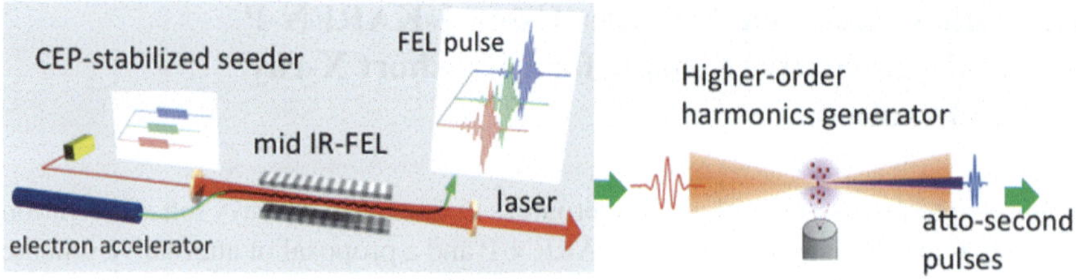

Fig. 6.4 Schematic figure of the concept of the higher order harmonics generator

by 15 J, 45 fs-duration pulse in the angle of incidence of 45°. The spot size is 2 μm-diameter, and the intensity is estimated to be 3×10^{21} W/cm^2, the pulse contrast in the 200 ps range is around 10^{-11}, whereas the pedestal or pre-pulses in several tens picosecond range still exist because we do not employ the plasma mirror system yet. Carbon ions are originated from contamination on the surface. Since J-KAREN-P is over-spec in terms of generating this energy range, we are currently developing a smaller size but higher repetitive (10–100 Hz) laser as shown in Chap. 2. By this new driver laser together with optimizations such as the use of pure carbon target, cleaning of the pre-pulses, we expect the required performance as the ion injector can be achieved.

Another topic is a proposal of alternative scheme toward high average power attosecond X-ray generation. Recent progress of atomic HHG extends the wavelength region into virtually keV region by use of near IR pumping lasers. Such soft X-ray sources are strongly desired from the field studying the laser interaction with atoms and molecules and the control of the chemical reaction. However, it can be said that the present conversion efficiency is not high so far, and the repetition rate does not reach the required level from pump-probe coincidence experiments, although the laser technologies are developing rapidly.

Our proposal is to use mid-infrared free-electron laser (mid IR-FEL) as an amplifier to realize MHz-class repetition rate [6]. An FEL oscillator with a perfect synchronized optical cavity is under consideration. Temporal shape of the FEL pulse consists of an exponential lobe of the leading edge and a main peak followed by ringing. Without seeding light, the amplified FEL pulse is generated from incoherent shot noise in the leading part, whose intensity is an order of 10^{-11} of that of the main peak. Under such condition, the amplitude and phase in the exponential envelope, the main peak, and the ringing are all affected by the interaction between the electrons and the radiation initiated by the shot noise in the leading part. As a result, the carrier frequency and phase of the FEL pulses are not stabilized. Whereas if we use a CEP-stabilized seed laser pulse whose intensity is an order of 10^{-7} with respect to that of the main peak, our simulation result indicates that the seed stabilizes the FEL pulse laser efficiently, and the original CEP is conserved. This result implies that if CEP-stabilized ultrashort pulses with the energy of nJ are injected into mid-IR-FEL, we can obtain CEP-stabilized 10 mJ-class mid IR pulses.

Figure 6.4 shows a schematic of the concept of the higher order harmonics generator, which consists of MHz-class CEP-stabilized seed laser, mid IR-FEL, and HHG part. Since all the components are technically well-established, we can expect 10 mJ-class MHz CEP-stabilized mid-IR pulses leading to 10 nJ-class MHz attosecond water window lasers, where we assume the conversion efficiency of 10^{-6}.

6.5 Summary

We showed recent progress and future prospect in the source development of SXRLs in transient collisional excitation (TCE) scheme with high repetition-rate operation and shorter wavelength HHG from plasma by use of our PW-class laser, J-KAREN-P. We also described research activities on laser-driven heavy ion acceleration for hadron therapy and high average power attosecond X-rays generation by using a combination of conventional electron accelerator and CEP-stabilized laser.

Acknowledgements This work has been done under the auspices of MIRAI-Program of Japan Science and Technology Agency (JST), Q-LEAP program of MEXT and Cross-ministerial Strategic Innovation Promotion Program (SIP) of Japanese Cabinet Office.

References

1. https://www.youtube.com/watch?v=Q5nVigMtino
2. Y. Ochi, N. Hasegawa, T. Kawachi, K. Nagashima, Development of a chirped pulse amplification laser with zigzag slab Nd:glass amplifiers dedicated to x-ray laser research. Appl. Opt. **46**, 1500–1505 (2007)
3. Y. Mikami, N. Hasegawa, H. Okada, S. Kondo, M. Nishikino, T. Kawachi, Flash-lamp-pumped 4 J, 50 Hz Nd:YAG nanosecond laser system for mobile and transportable equipment. Jpn. J. App. Phys. **56**, 082701 (2017)
4. A.S. Pirozhkov et al., Burst intensification by singularity emitting radiation in multi-stream flows. Sci. Rep. 7, 17968 (2017)
5. H. Kiriyama et al., High-contrast, high-intensity Petawatt-Class Laser and applications. IEEE Sel. Topics J. Quantum Electron **21**, 1601118 (2015)
6. R. Hajima, R. Nagai, Generating carrier-envelope-phase stabilized few-cycle pulses from a free-electron laser oscillator. Phys. Rev. Lett. **119**, 204802 (2017)

Chapter 7
Shaping of X-ray Pulses via Dynamical Control of Their Interaction with a Resonant Medium

T. Akhmedzhanov, V. Antonov, X. Zhang, K. C. Han, E. Kuznetsova, I. Khairulin, Y. Radeonychev, M. Scully and O. Kocharovskaya

Abstract High-harmonic generation sources allow producing trains of attosecond X-ray pulses in a water window range with a relatively small number of photons. We suggest a mechanism to amplify these trains of pulses in an inverted plasma of quasi-resonant ions driven by a moderately strong infrared (IR) laser field. The IR field modulates active medium with a double IR field frequency allowing for amplification of the X-ray field not only on the resonant transition frequency but also on sideband frequencies separated from the resonance by even number of modulation frequency. In the case of relatively strong plasma dispersion, each spectral component is amplified nearly independently. This allows amplifying X-ray radiation preserving it in the form of a train of attosecond pulses. We propose an experimental realization of the suggested technique in the active medium of C VI ions modulated by the IR laser field.

7.1 Introduction

Ultrashort coherent X-ray pulses provide a unique combination of spatial and temporal resolution, which might find numerous applications in applied and fundamental research, ranging from element-specific spectroscopy and controlling chemical reactions to manipulation of absorption and ionization properties of atoms on sub-femtosecond (sub-fs) time scales. One of the most promising applications of such pulses is an ultrafast imaging of large biological molecules, in particular, proteins. However, it typically requires sub-fs pulses containing a relatively large

---

T. Akhmedzhanov · X. Zhang · K. C. Han · E. Kuznetsova · M. Scully · O. Kocharovskaya (✉)
Department of Physics and Astronomy, Texas A&M University, College Station, TX 77843-4242, USA
e-mail: kochar@physics.tamu.edu

V. Antonov · I. Khairulin · Y. Radeonychev
Institute of Applied Physics, Russian Academy of Sciences, Nizhny Novgorod 603950, Russia

V. Antonov
Prokhorov General Physics Institute, Russian Academy of Sciences, Moscow 119991, Russia

M. Kozlová and J. Nejdl (eds.), *X-Ray Lasers 2018*, Springer Proceedings in Physics 241, https://doi.org/10.1007/978-3-030-35453-4_7

number of photons with a carrier frequency in the so-called water window range (4.4–2.3 nm–between the C and O K-shell absorption edges).

There are three conventional sources of ultrashort coherent X-ray pulses: high-harmonic generation (HHG) sources, X-ray free-electron lasers (XFELs) and X-ray plasma-based lasers. HHG allows producing pulses with down to tens of attoseconds duration in a table-top setup [1]. However, the X-ray range pulses usually have a relatively low number of photons (typical energy is on the order of few nJ in a pulse train). XFELs can produce few-fs pulses. However, these pulses suffer from the lack of temporal coherence coming from shot noise, which is an inevitable part in FEL amplification process. Besides, XFELs are large scale state-of-the-art expensive facilities and there are only a few of them in the world. X-ray plasma-based lasers are able of producing pulses with energies up to several mJ but of picosecond duration. Such pulses are too long for ultrafast imaging.

In this paper, we suggest a technique to amplify a train of sub-fs X-ray pulses produced by HHG via propagation through the inverted quasi-resonant plasma of hydrogen-like ions, simultaneously irradiated by a far-off-resonant IR laser field. In the presence of the IR field, the frequency of the resonant transition of the ions is modulated due to the linear Stark effect with a period, equal to half-period of the IR field. Due to this, the inverted medium amplifies the X-ray field not only at the frequency of the resonance, but also at its sidebands separated from each other by the doubled frequency of the IR field.

Suppose a train of X-ray pulses has carrier frequency exactly in resonance with the time-averaged frequency of the modulated transition of the active medium and its spectrum contains a number of harmonics separated by twice the modulating IR field frequency. Experimentally, such set up could be achieved if a replica of the IR field, which is used for HHG, modulates the plasma medium. In this scenario, all the spectral components of the incident X-ray field would be amplified during their propagation through the medium. For the plasma parameters considered in this paper, plasma dispersion is very strong for the IR field, which effectively cancels rescattering of spectral components of the X-ray field into each other. With the proper choice of the parameters of the modulating field, it allows to amplify the pulses preserving their temporal and spectral structure.

7.2 Propagation of X-ray Field Through IR-Field-Dressed Plasma Medium of Inverted Quasi-resonant Hydrogen-Like Ions

Let us assume now that at the entrance to the medium, an X-ray radiation has the form

$$\vec{E}_{\text{X-ray,inc}}(t) = \frac{1}{2}\vec{z}_0\tilde{E}_{z,0}(t)\exp\{-i\,\omega_z t\}+\frac{1}{2}\vec{y}_0\tilde{E}_{y,0}(t)\exp\{-i\,\omega_y t\} + \text{c.c.}, \quad (7.1)$$

where $E_{z,0}$ and $E_{y,0}$ are the slowly varying amplitudes of z-polarized and y-polarized components of the incident radiation, ω_z and ω_y are their carrier frequencies, and c.c. stands for complex conjugation. The medium consists of hydrogen-like ions with nuclear charge *–Ze*, where *Z* is atomic number and *e* is electron charge. The state of the ions is characterized by principal, *n*, orbital, *l*, and magnetic, *m*, quantum numbers. The radiation (1) is chosen to be near-resonant to a transition from the first excited energy level of the ions, $n = 2$, to the ground energy level, $n = 1$, that is $\omega_z \approx \omega_y \approx \omega_{21}^0$, where ω_{21}^0 is the unperturbed frequency of the resonant transition.

Propagation of X-ray radiation through the medium is described by the wave equation

$$\frac{\partial^2 \vec{E}_{\text{X - ray}}}{\partial x^2} - \frac{\varepsilon}{c^2}\frac{\partial^2 \vec{E}_{\text{X - ray}}}{\partial t^2} = \frac{4\pi}{c^2}\frac{\partial^2 \vec{P}}{\partial t^2}, \tag{7.2}$$

where $\vec{E}_{\text{X - ray}} = \vec{z}_0 E_z + \vec{y}_0 E_y$ is the X-ray field strength, $\vec{P}$ is the resonant polarization of the medium. We take into account now the following (lasing) states of hydrogen-like ions: the ground state $|1s\rangle$, denoted as $|1\rangle$, and four excited states: $|2\rangle = (|2s\rangle + |2p, m = 0\rangle)/\sqrt{2}$, $|3\rangle = (|2s\rangle - |2p, m = 0\rangle)/\sqrt{2}$, $|4\rangle = |2p, m = 1\rangle$, and $|5\rangle = |2p, m = -1\rangle$ which are degenerate in the absence of the IR field. The range of the IR field parameters (frequency and amplitude) where this approximation is valid was studied in detail in [2]. Then, the resonant polarization is defined by density matrix elements ρ_{ij}:

$$\vec{P}(\vec{r}, t) = N\left(\vec{d}_{12}\rho_{21} + \vec{d}_{13}\rho_{31} + \vec{d}_{14}\rho_{41} + \vec{d}_{15}\rho_{51} + \text{c.c.}\right), \tag{7.3}$$

where N is ion concentration and $\vec{d}_{ij}$ is the dipole moment of transition between states $|i\rangle$ and $|j\rangle$. The nonzero dipole moments include also

$$\begin{gathered}\vec{d}_{12} = \vec{z}_0 d_{tr}, \quad \vec{d}_{13} = -\vec{z}_0 d_{tr}, \quad \vec{d}_{14} = i\vec{y}_0 d_{tr}, \quad \vec{d}_{15} = i\vec{y}_0 d_{tr}, \\ \vec{d}_{22} = \vec{z}_0 d_{av}, \quad \vec{d}_{33} = -\vec{z}_0 d_{av}, \quad \vec{d}_{ij} = \vec{d}_{ji}^*,\end{gathered} \tag{7.4}$$

where $d_{tr} = d_{1s\leftrightarrow 2p}/\sqrt{2} = 2^7/(3^5 Z)$ and $d_{av} = d_{2s\leftrightarrow 2p} = 3/Z$, (hereinafter atomic units are assumed throughout the paper). Under the action of both X-ray and IR field, the evolution of density matrix elements is given by the following equations:

$$\begin{gathered}\dot{\rho}_{11} = +\gamma_{11}(\rho_{22} + \rho_{33} + \rho_{44} + \rho_{55}) - i[\text{H}, \rho]_{11} \\ \dot{\rho}_{ij} = -\gamma_{ij}\rho_{ij} - i[\text{H}, \rho]_{ij}, ij \neq 11 \\ \text{H} = \begin{pmatrix} \omega_1 & -d_{tr}E_z & d_{tr}E_z & -id_{tr}E_y & -id_{tr}E_y \\ -d_{tr}E_z & \omega_2 - d_{av}E_{IR}(x,t) & 0 & 0 & 0 \\ d_{tr}E_z & 0 & \omega_3 + d_{av}E_{IR}(x,t) & 0 & 0 \\ id_{tr}E_y & 0 & 0 & \omega_4 & 0 \\ id_{tr}E_y & 0 & 0 & 0 & \omega_5 \end{pmatrix}\end{gathered} \tag{7.5}$$

where $E_{IR}(x,t) = \tilde{E}_{IR}\cos\{-i\Omega(t - xn_{pl}/c)\}$ is the IR field strength, while ω_i is energy of the state $|i\rangle$. Equation (7.5) assumes that the IR field has the same linear polarization as the X-ray field. With quadratic Stark effect taken into account, $\omega_1 = -\frac{Z^2}{2}\left(1 + \frac{9}{256}F_c^2\right)$, $\omega_2 = \omega_3 = -\frac{Z^2}{8}\left(1 + \frac{21}{4}F_c^2\right)$, $\omega_4 = \omega_5 = -\frac{Z^2}{8}\left(1 + \frac{39}{8}F_C^2\right)$, where $F_c = (2/Z)^3\tilde{E}_{IR}$ is normalized IR field strength. The decay rates γ_{ij} are $\gamma_{12} = \gamma_{13} = \gamma_{\rm coll} + \Gamma_{\rm ion}^{(2,3)}/2 + \Gamma_{\rm rad}/2$, $\gamma_{14} = \gamma_{15} = \gamma_{\rm coll} + \Gamma_{\rm ion}^{(4,5)}/2 + \Gamma_{\rm rad}/2$, $\gamma_{23} = \gamma_{\rm coll} + \Gamma_{\rm ion}^{(2,3)} + \Gamma_{\rm rad}$, $\gamma_{45} = \gamma_{\rm coll} + \Gamma_{\rm ion}^{(4,5)} + \Gamma_{\rm rad}$, $\gamma_{24} = \gamma_{25} = \gamma_{34} = \gamma_{35} = \gamma_{\rm coll} + \Gamma_{\rm ion}^{(2,3)}/2 + \Gamma_{\rm ion}^{(4,5)}/2 + \Gamma_{\rm rad}$, $\gamma_{22} = \gamma_{33} = \Gamma_{\rm ion}^{(2,3)} + \Gamma_{\rm rad}$, $\gamma_{44} = \gamma_{55} = \Gamma_{\rm ion}^{(4,5)} + \Gamma_{\rm rad}$, $\gamma_{11} = \Gamma_{\rm rad}$. Here $\Gamma_{\rm rad}$ is radiative decay rate [3], $\gamma_{\rm coll}$ is collision rate [4], $\Gamma_{\rm ion}^{(2,3)}$ and $\Gamma_{\rm ion}^{(4,5)}$ are ionization rates from the states $|2\rangle$, $|3\rangle$, and $|4\rangle$, $|5\rangle$, correspondingly [5]. We neglect Doppler broadening of transition lines since for ion temperatures on the order of 1 eV, which are typical for recombination X-ray plasma lasers [6], Doppler broadening is comparable to or less than collisional broadening. For the parameters we consider in the paper, $\gamma_{ij} \ll \Omega$.

In the following, we assume the slowly varying amplitude approximation for both (i) the X-ray field and (ii) the quantum coherencies of transitions between the ground and excited states, as well as the rotating wave approximation at these transitions. We introduce the local time $\tau = t - x/c$, and assume as initial conditions that at $\tau = 0$ all the ions are pumped to the states $|2\rangle$, $|3\rangle$, $|4\rangle$, and $|5\rangle$ with equal probability, and there are no coherencies in the medium:

$$\rho_{ii}(x, \tau = 0) = 1/4,\ i = 2, 3, 4, 5; \quad \rho_{ij}(x, \tau = 0) = 0,\ ij \neq 22, 33, 44, 55 \tag{7.6}$$

Equation (7.6) implies that the population inversion is created by a running pump wave, which moves at the speed of light. As boundary conditions we assume $E_z(x = 0, \tau) = \tilde{E}_{z,0}(t)$ and $E_y(x = 0, \tau) = \tilde{E}_{y,0}(t)$.

As it can be seen from (7.5), the instantaneous frequencies of transitions $|2\rangle \leftrightarrow |1\rangle$ and $|3\rangle \leftrightarrow |1\rangle$ are periodically modulated with a period equal to that of the IR field. Physically it corresponds to linear AC Stark effect. Since transitions $|2\rangle \leftrightarrow |1\rangle$ and $|3\rangle \leftrightarrow |1\rangle$ have nonzero components of dipole moment along z-direction, the modulation of frequencies of these transitions will result in modulated response of the medium to the z-polarized X-ray field propagating in x-direction. Namely, a gain will appear at sideband frequencies separated from the frequency of the resonance by by even multiples of the frequency of the IR field. Consider z-polarized component of an incident X-ray field (1) to be a train of ultrashort pulses with the carrier frequency equal to the time-averaged values of frequencies of the transitions $|2\rangle \leftrightarrow |1\rangle$ and $|3\rangle \leftrightarrow |1\rangle$ (which account for the quadratic Stark effect and are the same for both transitions) and a pulse period equal to a half period of the modulating IR field. In this case, each spectral component of the incident X-ray field will be in resonance with the corresponding spectral component of the medium gain. In general, the gain differs for different spectral components of the incident field and its spectral distribution is determined by the parameters of the modulation field. With a proper choice of these parameters and the medium length, it appears to be possible to have the incident train of the sub-fs pulses amplified at the output of the medium.

7.3 Amplification of a Train of Sub-fs Pulses in the IR-Field-Dressed Plasma Medium of Inverted Quasi-resonant Hydrogen-Like Ions

Let us consider a neutral plasma consisting of C VI ions, electrons, and some other ions (for example, H^+ to maintain electric neutrality) with C VI ion concentration $N = 10^{19}$ cm^{-3} and electron density $N_{el} = 15$ N. Lasing in inverted plasma with such parameters has been theoretically studied [6] and is under the experimental investigation by the group of Prof. Szymon Suckewer in the University of Princeton, USA. Let such plasma is modulated by the IR field with wavelength $\lambda_{IR} = 2102.85$ nm and intensity $I_{IR} = 2.7 \times 10^{15}$ W/cm^2.The plasma refractive index for the IR field is $n_{pl} \approx 0.6$ and its 623th harmonic is in exact resonance with the transitions $|2\rangle \leftrightarrow |1\rangle$ and $|3\rangle \leftrightarrow |1\rangle$. Such modulating field corresponds to the so-called index of modulation $P_\omega = d_{av}\tilde{E}_{IR}/\Omega = 6.4$. The modulation index defines how many sidebands will have comparable amplitudes in the gain frequency distribution. The value of 6.4 corresponds to 3 sidebands of comparable amplitudes at each side from the resonance (in addition to the central spectral component of the gain).

Let an incident z-polarized X-ray field be a train of ultrashort pulses resonant to the transitions $|2\rangle \leftrightarrow |1\rangle$ and $|3\rangle \leftrightarrow |1\rangle$ with Gaussian envelope of the train: $\tilde{E}_{z,0}(t) = E_{z,0}^{\max} \exp\{-2.773(t - 2t_{\text{env}})^2/t_{\text{env}}^2\} \sum_{k=-3}^{k=3} \exp\{-2ik\Omega t - ik^2 a_{\text{chirp}}\}$, where $t_{\text{env}} = 20$ fs, $a_{\text{chirp}} = \pi/(2n_{\max})$, $n_{\max} = 763$, and the peak intensity of z-polarized incident X-ray field is chosen to be equal to $I_z(x = 0) = 10^6$ W/cm^2. Let us first assume that the incident y-polarized X-ray field is zero. The results of numerical solution of the (7.2)–(7.5) describing the propagation of such field through 1 mm of the plasma medium modulated by the IR field are shown in Fig. 7.1. As it can be seen, the train of the pulses was amplified: the peak intensity at the exit of the medium is about 48 times larger compared to the entrance. It is worth noting that contrary to the case of low-density plasma considered in [7], the sidebands of the incident X-ray field do not scatter into each other during their propagation through the modulated medium because of very small length of their mutual coherent interaction (originated from strong plasma dispersion at the frequency of the modulating IR field).

In Fig. 7.1, we have considered the propagation of the z-polarized pulse train through the amplifying medium assuming there is no y-polarized field. This assumption is not strictly valid since in the inverted medium an amplification of spontaneous emission with y-polarization will occur, leading to the formation of relatively strong y-polarized X-ray field. The peak value of the gain for such field will be several times larger than that for the z-polarized X-ray field (which is reduced because of redistribution of the gain between the sidebands under the action of z-polarized the modulating field). As a result of a presence of the strong y-polarized field, a part of

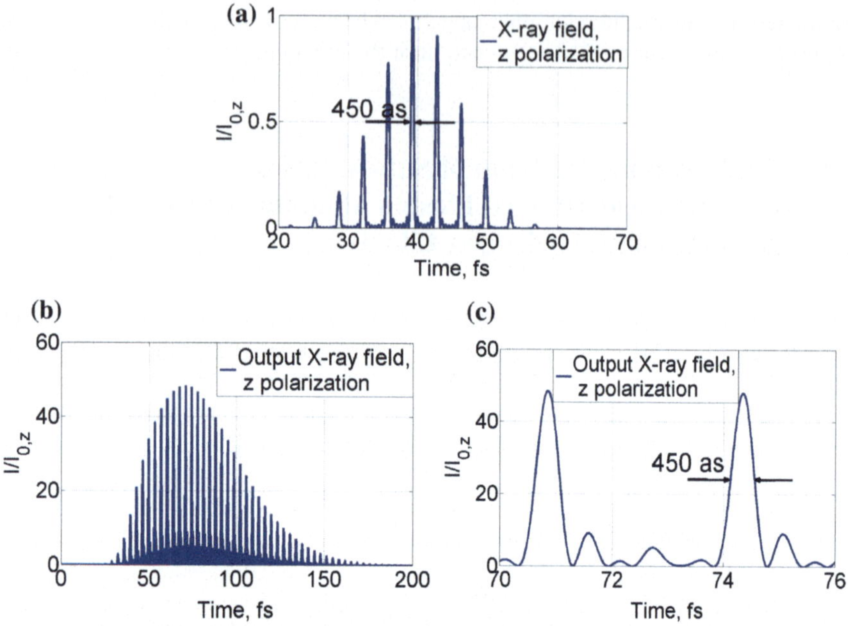

Fig. 7.1 **a** Incident z-polarized X-ray pulse train; **b** z-polarized X-ray pulse train at the output of 1 mm of modulated inverted C^{5+} plasma medium for $I_y(x = 0) = 0$; **c** the same as (**b**), but with higher temporal resolution. The total energy of the output z-polarized X-ray field is 232 times larger than that of the input z-polarized X-ray field

the population of excited states $|4\rangle$ and $|5\rangle$ will drop to ground state $|1\rangle$ so that the population inversion on the transitions $|2\rangle \leftrightarrow |1\rangle$ and $|3\rangle \leftrightarrow |1\rangle$ as well as gain for the z-polarized field at these transitions will be decreased. In order to investigate this effect, we numerically solved the Eqs.(7.2)–(7.5) for the same parameters as in the Fig. 7.1, but with non-zero y-polarized incident X-ray field (which we assumed to be monochromatic and resonant to the transitions $|4\rangle \leftrightarrow |1\rangle$ and $|5\rangle \leftrightarrow |1\rangle$). The results are shown in Figs. 7.2 and 7.3 for the two values of intensity of this incident y-polarized field: I_y $(x = 0) = 10^3$ W/cm^2 and I_y $(x = 0) = 10^6$ W/cm^2. As can be seen, the incident y-polarized field is strongly amplified in the medium. It leads to lower gain for z-polarized X-ray field and lower peak intensities and energies of the amplified pulse train. Nevertheless, the amplification clearly occurs, and the pulses maintain their duration of 450 as even in the case of I_y $(x = 0) = I_z(x = 0)$.

7.4 Conclusion

In this paper, we have shown the possibility to amplify a train of sub-fs X-ray pulses in the medium of hydrogen-like ions dressed by a strong IR field. The IR field modulates

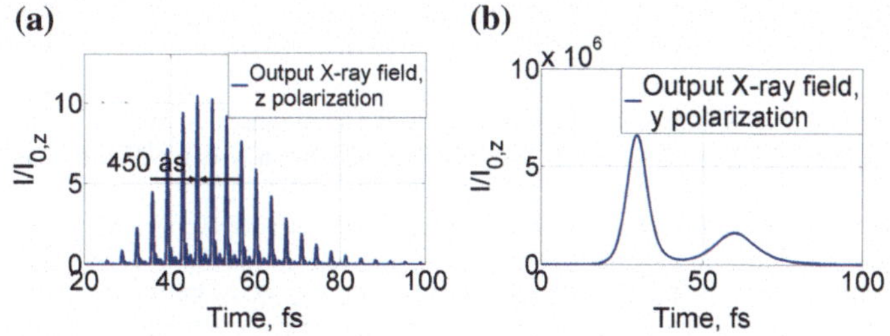

Fig. 7.2 **a** z-polarized X-ray pulse train at the output of 1 mm of modulated inverted C^{5+} plasma medium for $I_y(x = 0) = 10^3$ W/cm^2; **b** y-polarized X-ray radiation at the output of 1 mm of modulated inverted C^{5+} plasma medium for $I_y(x = 0) = 10^3$ W/cm^2. The total energy of z-polarized X-ray radiation is 20 times larger than that of the input z-polarized X-ray radiation

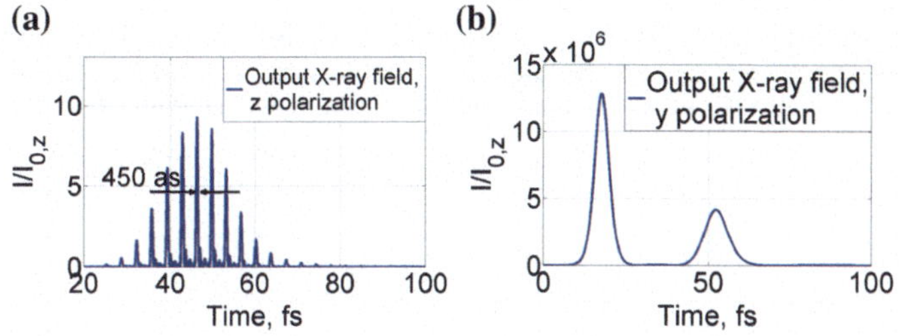

Fig. 7.3 The same as in Fig. 7.2, but for $I_y(x = 0) = 10^6$ W/cm^2. The output z-polarized X-ray field is amplified 12.5 times with respect to the incident one (in terms of the total energy)

the resonant X-ray transition of the ions and leads to the appearance of sidebands of the medium gain, which allows amplification of all spectral components of the incident attosecond pulse train. The possibility to amplify a train of 450 as pulses with the central wavelength in the vicinity of 3.4 nm (in the "water window" range) in the medium of C VI ions dressed by the 2.1 μm IR field was shown. In such setup, the output X-ray field is also a train of 450 as pulses, but with much larger energy and peak intensity. The suggested technique of X-ray sub-fs pulse amplification might be useful for ultrafast imaging of biological samples with a high special resolution.

Acknowledgements This work was supported by the NSF grant No. PHY-1596467; the King Abdulaziz City for Science and Technology (KACST) program, Robert A. Welch Foundation grant No. A-1261, as well as by RFBR under the grant No. 18-02-00924. The studies presented in Sect. 7.2 were supported by the Ministry of Education and Science of the Russian Federation under contract No.14.W03.31.0032. V. A. A. acknowledges a personal grant from Foundation for the Advancement of Theoretical Physics and Mathematics "BASIS".

References

1. J. Li, X. Ren, Y. Yin, K. Zhao, A. Chew, Y. Cheng, E. Cunningham, Y. Wang, S. Hu, Y. Wu, M. Chini, Z. Chang, 53-attosecond X-ray pulses reach the carbon K-edge. Nat. Commun. **8**, 186 (2017)
2. T.R. Akhmedzhanov, MYu. Emelin, V.A. Antonov, Y.V. Radeonychev, MYu. Ryabikin, O. Kocharovskaya, Ultimate capabilities for few-cycle pulse formation via resonant interaction of XUV radiation with IR-field-dressed atoms. Phys. Rev. A **95**, 023845 (2017)
3. A. Kramida, Y. Ralchenko, J. Reader, NIST ASD Team (2015), (NIST Atomic Spectra Database (ver. 5.3). http://physics.nist.gov/asd [2017, January 1], National Institute of Standards and Technology, Gaithersburg, MD
4. H.R. Griem, *Spectral Line Broadening by Plasmas* (Acad. Press, New York, 1974)
5. V.S. Popov, Tunnel and multiphoton ionization of atoms and ions in a strong laser field (Keldysh theory). Phys.-Usp. **47**, 855 (2004)
6. Y. Avitzour, S. Suckewer, Feasibility of achieving gain in transition to the ground state of C VI at 3.4 nm, J. Opt. Soc. Am. B **24**, 819 (2007)
7. T.R. Akhmedzhanov, V.A. Antonov, A. Morozov, A. Goltsov, M. Scully, S. Suckewer, O. Kocharovskaya, Formation and amplification of sub-femtosecond x-ray pulses in a plasma medium of hydrogenlike ions with a modulated resonant transition. Phys. Rev. A **96**, 033825 (2017)

Chapter 8
Seeded Soft X-Ray Lasers: Towards OAM Beam Amplification and Quasi-Pi-pulse Regime

O. Guilbaud, F. Sanson, A. K. Pandey, E. Papagiannouli, J. Demailly, O. Neveu, B. Lucas, E. Baynard, M. Pittman, D. Ros, F. Harms, G. Dovillaire, M. Richardson, Ph. Balcou and S. Kazamias

Abstract We present in this paper ongoing studies on seeded plasma-based soft X-ray lasers using the LASERIX facility. After an overview of the source setup and characteristics, we will discuss two prospects that might increase the interest of these coherent soft X-ray sources. We have successfully generated harmonic beams carrying Orbital Angular momentum. We briefly present this experiment and discuss the remaining challenges for amplifying these kinds of beams in a plasma-based soft X-ray laser. In the second part, we present numerical results on harmonic seed amplification in conditions leading to Rabi oscillations. This regime is known to be a promising way to reach femtosecond pulses in the present plasma amplifiers. We will discuss the possibility of increasing the stability of such a regime by the introduction of a continuous refractive loss in the gain region.

O. Guilbaud (✉) · F. Sanson · A. K. Pandey · E. Papagiannouli · J. Demailly · O. Neveu · B. Lucas · D. Ros · S. Kazamias
Laboratoire de Physique de Gaz et de Plasmas, Universite Paris-Saclay (UMR-CNRS 8578), Bat. 210, 91405 Orsay Cedex, France
e-mail: olivier.guilbaud@u-psud.fr

E. Baynard · M. Pittman
LUMAT, Universite Paris-Saclay (UMR-CNRS 8578), Bat. 210, 91405 Orsay Cedex, France

F. Harms · G. Dovillaire
Imagine Optic, 18, rue Charles de Gaulle, 91400 Orsay, France

M. Richardson
University of Central Florida, 4304 Scorpius Street, Orlando, FL 32816-2700, USA

M. Richardson · Ph. Balcou
Centre Lasers Intenses et Applications, Universite de Bordeaux (UMR-CNRS-CEA 5107), Talence, France

M. Kozlová and J. Nejdl (eds.), *X-Ray Lasers 2018*, Springer Proceedings in Physics 241, https://doi.org/10.1007/978-3-030-35453-4_8

8.1 Introduction

The LASERIX facility is an experimental platform of the Paris-Saclay University. The core of this facility is a titanium–sapphire laser delivering 2 J, 40 fs infrared pulses at a 10 Hz repetition rate. As will be discussed later, the main part of the energy beam is usually dedicated to the generation of coherent soft X-ray lasers (SXRL). However, the facility is evolving to offer new capabilities and covering new domains. Between the laser front-end and the main amplifiers, small fractions of the laser energy are diverted, recompressed to 30 fs and delivered to new experimental points. One of these auxillary beams is used for the development of a Sagnac-type interferometer. This device is part of an ongoing project to measure the QED deflexion of a probing light beam by an intense light beam in vacuum [1].

8.2 Seeded Operation of a Solid Target Soft X-Ray Laser

A general overview of the X-ray laser beamline is presented in Fig. 8.1. The plasma amplifier stage is a plasma in which a population inversion is induced on the 3p-3s (J = 0 − 1) transition ($\lambda = 32.6$ nm) of Neon-like Titanium ions by electron collisions. As described in [2] the plasma amplifier is formed by the interaction of two laser pulses with a Titanium solid slab target. A first pulse (300 mJ with 8 ns duration) from a Q-switched Nd-YAG laser, is forming an expanded, weakly ionized plasma. After an optimum 4 ns peak-to-peak delay, a 900 mJ compressed pulse from the main Ti:Sa laser, is line-focused on the plasma to generate the SXRL gain. The grazing incidence focusing configuration [3] leads to a 4 mm $\times$ 15 μm focal spot on target with a travelling wave pumping speed close to the speed of light. The laser energy is absorbed by the plasma at an electron density close to $N_e = N_c sin^2(\Phi)$ where $\Phi = 22°$ is the grazing angle and N_c is the infrared laser critical density. An optimal

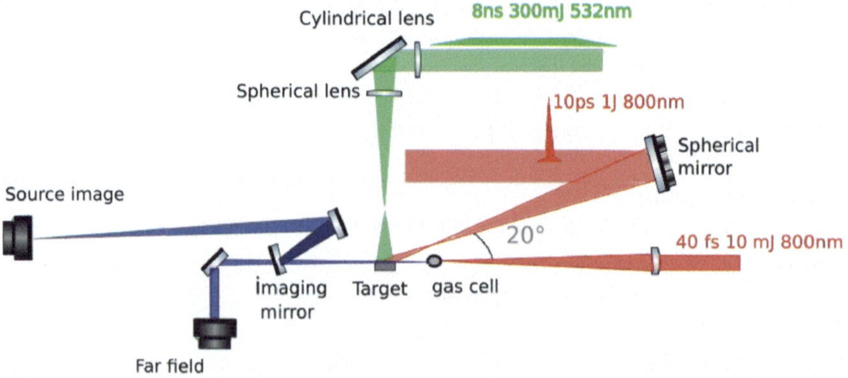

Fig. 8.1 Overview of the LASERIX seeded soft X-ray laser system

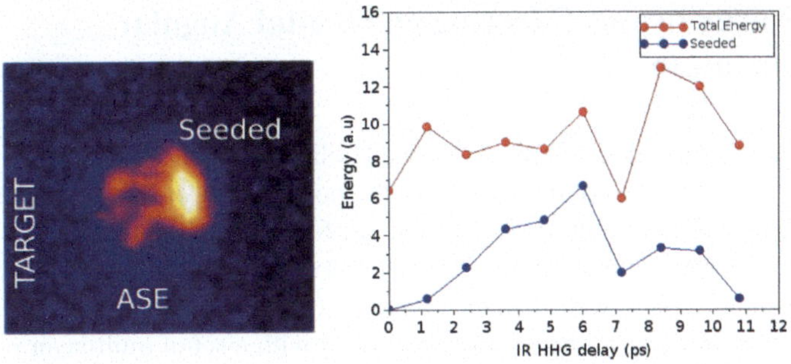

Fig. 8.2 Contribution of ASE and seeded signal to the total soft X-ray laser energy output. Left: near field image of the Ne-like Titanium soft X-ray laser ($\lambda = 32.6$ nm). The target is located on the left of the image. Seeded part of the beam and ASE contribution are highlighted. Right: evolution of the energy in the seeded part and of the total energy as a function of injection time

soft X-ray output energy has been found for a 5 ps short pulse duration. The soft X-ray laser beam is monitored with a spectrometer (not represented on the scheme), a footprint camera placed 1 m away from the source. Besides the source is imaged on a soft x-ray camera by multilayer spherical mirror with a magnification of ten.

The amplifier stage is seeded with the 25th harmonic of the infrared laser [4]. High order harmonics are generated using a low energy ($\leq$10 mJ) auxiliary beam compressed to 40 fs and focused by an f = 1 m lens into a 10 mm long gas cell. This cell is filled with 20 mbar of Argon and is placed 20 mm before the plasma amplifier. The Harmonic spectral tuning is achieved via IR spectral phase control through an acousto-optical modulator (Dazzler). In this direct seed amplification, in which no relay optics is used to couple the seed to the amplifier, the seed is observed to be two times larger than the soft x-ray laser emitting region ($20\,\mu m \times 40\,\mu m$). As depicted in Fig. 8.2, the contribution of the seeded beam exiting the plasma and the contribution of the ASE can be identified in the images of the source. With an injection angle close to 0 and a 4 mm long plasma, the seeded emission is always appearing at a higher distance from the target than the ASE emission. The seeded beam is probably refracted by the plasma density gradient during its propagation. This is consistent with a 6 mrad deflexion angle observed in far field. Seeded signal is observed during a 8 ps window of seed injection time. In this experimental condition, the total energy of seeded laser plus ASE emission is found almost constant with the injection time. At the optimal delay value, 60 percent of the lasing energy is in the seeded beam.

8.3 Towards Beams Carrying an Orbital Angular Momentum

Different groups have demonstrated that seed polarization is preserved by the amplification process [5]. The possibility to turn the polarization of the seeded laser beam is implemented in our setup with a simple half-wave plate placed on the infrared laser beam generating the harmonics. Furthermore, powerful plasma-based soft X-ray laser beams with a well-defined circular polarization have been achieved by conferring this property to the harmonic seed [6] with a set of multilayer optics. An interesting perspective would be to extend this approach to beam carrying Orbital Angular Momentum (OAM). Different groups have already demonstrated the possibility of imprinting this topological property of light wavefront to the high-order harmonics beams [7, 8]. The amplification of an OAM beam by a plasma-based soft X-ray laser might open new applications involving tight focusing of the EUV beam, such as coherent diffraction imaging, or ptychography. Our seeded laser setup has been modified to enable such an experiment. However, many aspects need to be addressed. Among them, a reliable diagnostics demonstrating that a "donut shape" beam is indeed carrying orbital momentum is required.

The first step of this project has been recently achieved [9]. We have demonstrated the efficient generation of OAM harmonic beams at a wavelength corresponding to the Ne-like Titanium laser. A $l = +1$ orbital angular momentum is first conferred to the infrared harmonic generating beam by inserting a phase plate on the beam path before the focusing lens. The phase plate is composed of 16 sectors of different thicknesses and is adding to the incident beam a spatial phase ϕ approximately equals to $\phi = l\theta$, θ being the azimuthal angle and l being an integer. For this particular plate, the value of l was one. The harmonic beam is c filtered by a plan multilayer mirror selecting the 25th harmonics (32 nm). The beam wavefront is finally analysed using an EUV Hartmann wavefront sensor developed by the Imagine Optics company. As depicted in Fig. 8.3, the far-field harmonic beam is formed of an incomplete ring, with a divergence higher than the divergence of classical harmonics. This partial ring

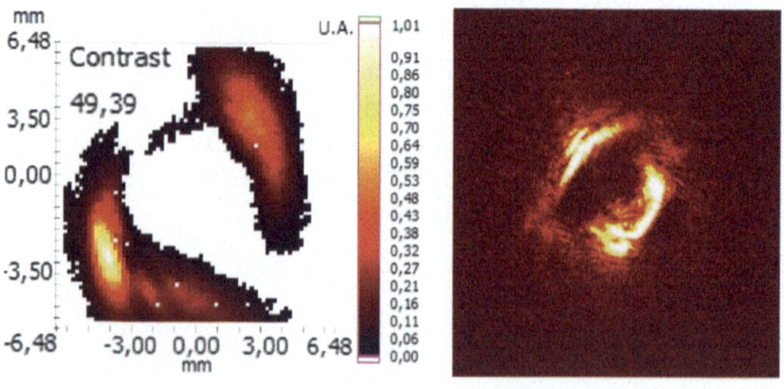

Fig. 8.3 Far-field and near-field images of the 25th harmonic beam carrying orbital angular momentum

structure is also observed in near field using the soft X-ray imager described above. The wavefront analysis enabled us to find a dominating wavefront component carrying an orbital momentum with a charge $l = 25$. It should be noted that these beams were generated in a long medium, in conditions associated with phase-matching for classical harmonics. The origin of the incomplete ring has been identified during a second experiment. The correction of a remaining astigmatism in the IR generating beam leads to the generation of a complete annulus. We have also observed that the OAM beam has the same polarization as the infrared generating field. Besides, the signal level obtained for this OAM beams has the same order of magnitude as classical harmonics. This is an encouraging result for soft x-ray laser amplification project. However, the large divergence of the harmonic beam will make the harmonic coupling to the gain region more challenging than expected. Refocusing of the seed beam or increasing the plasma amplifier gain region are both options under investigation.

8.4 Quasi-pi-Pulse Operation

The pulse duration of the isolated harmonic seed (<30 fs) is two orders of magnitude smaller than the two characteristics times of the amplifier, namely its dephasing time T_2, which is also the inverse of the homogeneous bandwidth, and the recovery time T_1 which is of the order of few ps [10]. Despite its low energy, the seed has a peak intensity close to the saturation intensity I_{sat} of the amplifier. These particular features, associated with very high gain values ($g_0 \sim 100\,\text{cm}^{-1}$), is predicted to lead to a complex behaviour, described by the non-adiabatic Bloch–Maxwell equations. The theoretical evolution of the SXRL pulse during its amplification has been clearly established [11–13]. The harmonic seed is triggering a coherent polarization in the medium with a response time of the order of T_2. A picosecond XUV wake is formed after the harmonic seed, becoming more and more intense as the pulse is propagating along the amplifier. When the saturation intensity is reached, Rabi oscillations will form at the beginning of the wake. The frequency of these oscillations will increase with the amplification length and their peak intensity can become one to two orders of magnitude higher than the saturation intensity. Therefore, a high intensity, femtosecond pulse can, theoretically, be supported by a classical plasma-based X-ray laser. Many issues need to be overcome before observing and exploiting this dramatic effect. The amplification length shall be sufficiently long for the pulse to reach the oscillation regime with very high Rabi frequency. The gain medium has to be uniform in order to get the same temporal profile along the section of the beam and thus avoiding a smearing of the oscillations. Besides, a picosecond long and low intensity wake is following the Rabi main pulse, containing a substantial fraction of the pulse. Finally, in these conditions, the ASE may hide the phenomenon and will even enter in competition with its formation.

An interesting way to make this regime robust is to introduce a controlled amount of linear loss in the gain medium. This loss will force the pulse shape to converge towards a single pulse with a stable profile [14]. In the particular case of a long-lasting pumping rate, a succession of pulses separated by a time interval close to the recovery time T_1 will be obtained. The population inversion just after each pulse is fully inverted. For this reason, such kind of pulse is named a quasi-pi-pulse. Their duration is proportional to the ratio of the loss rate to the population inversion. In a soft x-ray laser, the loss can be the inverse Bremstrahlung absorption by plasma electrons. However, this loss is so low that the steady-state regime would be achieved after tens of centimeters. It has been proposed by Larroche et al. [15], that in realistic soft X-ray laser, refraction and diffraction can act as losses for the beam travelling through the gain region. However, for the global beam, this energy is not lost but diverted away from the propagation axis.

The plasma structure explored by Larroche is a localized gain region in a quasi exponential density gradient. We developed further this idea through numerical simulation, in which we consider a simplified plasma amplifier configuration. As represented by the yellow curve in Fig. 8.4 (bottom), the electron density transverse profile is a Gaussian electron density bump on top of a uniform electron density. Centred on this bump, a Gaussian-shape gain region is assumed with a peak small signal gain of 200 cm^{-1} (red curves in Fig. 8.4. The gain is supposed to be governed by a pumping rate with a Gaussian temporal envelope of 4ps FWHM. The recovery time T_1 is set to 2 ps, whereas the dephasing time T_2 is supposed to be inversely proportional to N_e, with a value of $T_2 = 500$ fs for $N_e = 2 \times 10^{20}$ cm^{-3}. The seed is injected at the peak of the pump rate. Its pulse duration is 40 fs, for a peak intensity of $0.1 I_{sat}$ and a waist two times larger than the density bump. This refractive amplifier configuration is compared to a reference situation in which the bump is supposed to be very large compared to the harmonic waist, with a super Gaussian profile leading to an almost uniform density profile (top of Fig. 8.4).

Simulated field in the amplifier at a particular time is presented in the form of color maps in Fig. 8.4. In the non-refractive case (top of Fig. 8.4), a longitudinal profile similar to the one observed in 1D Bloch–Maxwell simulations is observed [11]. The leading intense pulse resulting from the first Rabi oscillation is followed by a longer wake. After 1 mm of propagation, the transverse size of the beam converges to the gain region size. In the refractive case (top of Fig. 8.4), an isolated pulse separated from a second one by an interval close to cT_1 is observed. It should be emphasized that no signal is observed between these pulses. The main pulse duration is ~160 fs which is smaller than T_2. For amplifiers longer than 4 mm, the longitudinal structure is not significantly changing. As an example, the pulse duration and intensity are the same for a 10 and 20 mm amplifiers. However, the size of the beam is increasing linearly with the amplification length. Besides, the peak intensity reached in the refractive case is lower than in the non-refractive case, but the spatially integrated power is similar to the integrated power of the non refractive case. The energy lost in the gain region is indeed deflected away by refraction, feeding the wings of the beam.

A preliminary study has been performed to evaluate the impact of the ASE on both regimes. In the refractive regime, if an intense leading soliton-pulse can form, it will inhibit the formation of ASE in the wake. However, if the population inversion

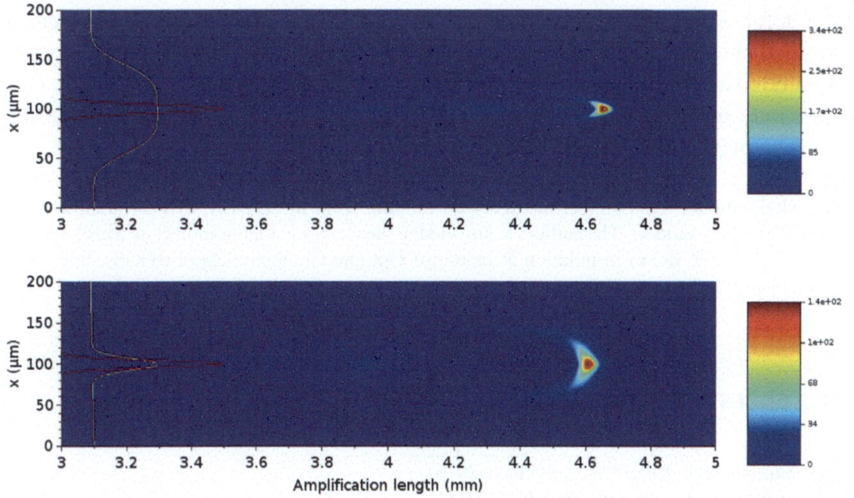

Fig. 8.4 2D Bloch–Maxwell simulation of an harmonic seed amplification into a plasma amplifier. On Top: amplifier presenting a uniform transverse density profile. Bottom, the amplifier presents a density bump centred at $x = 100\,\mu\text{m}$. The intensity color scale is expressed in unit of I_{sat}. The red and yellow curves represent the gain and electron density profiles, respectively, with a peak gain value of $200\,\text{cm}^{-1}$ and a peak density value of $3 \times 10^{20}\,\text{cm}^{-3}$

is established many picosecond prior to the seed, an ASE pedestal can grow sufficiently to provoke local disruption of the soliton-pulse front. The density and gain profile proposed here to reach this regime might be accessible experimentally. Plasma hydrodynamic simulation of our soft X-ray laser pumping configuration shows the possible existence of a density bump where the gain is forming. This density bump is forming during the interaction of the picosecond pumping pulse with the cold weakly ionized large pre-plasma. Optimization of the gain centering and density bump symmetry might be possible by varying the grip angle, the pumping pulse duration, or by introducing two pumping pulses.

8.5 Conclusion

We present in this proceeding some possible perspective of seeded operation of a plasma-based soft X-ray laser. We are presently working actively on the amplification of an harmonic seed carrying Orbital Angular Momentum. The generation of such a seed is now implemented and demonstrated in our experimental setup. Because the seed beam divergence is strongly increasing, the main challenge remains the spatial coupling of the seed with the ASE. Additional harmonic focusing is under consideration. Furthermore, we investigate theoretically the influence of some particular density profiles in the evolution of the seeded beam spatio-temporal shape. We show that a defocussing profile placed in the gain region can act as a linear loss that, if controlled can lead to the formation of a quasi-pi-pulse. Peak intensity is decreased compared to a refraction-less situation but the spatially integrated power is not. This regime may be a robust way to reach isolated femtosecond Rabi oscillation at the exit of the plasma-based soft X-ray laser.

Acknowledgements This project has received funding from the LIDEX OPT2X grant and the Sesame PULSE-X support. The authors acknowledge the Centrale d'laboration et de Mtrologie des Optiques XUV (CEMOX) installation at Institut d'Optique Graduate School (IOGS), Palaiseau, France, for providing the multilayer optics.

References

1. X. Sarazin et al., Eur. Phys. J. D **70**, 1 (2016)
2. O. Delmas et al., Opt. Lett. **39**, 6102 (2014)
3. R. Keenan et al., Phys. Rev. Lett. **94**, 103901 (2005)
4. O. Guilbaud et al., Opt. lett. **40**, 4775–4778 (2015)
5. Ph Goddet et al., Opt. Lett. **32**, 1498 (2007)
6. A. Depresseux et al., Phys. Rev. Lett. **115**, 083901 (2015)
7. C. Hernandez-Garcia et al., Phys. Rev. Lett. **111** 083602 (2013)
8. R. Geneaux et al., Nat. Commun. **7**, 12583 (2016)
9. F. Sanson et al., Opt. Lett. **43**, 2780–2783 (2018)
10. Y. Wang et al., Nat. Photonics **8**, 381–384 (2014)
11. C.M. Kim et al., Phys. Rev. A **80**, 053811 (2009)
12. I.R. Al'miev et al., Phys. Rev. Lett. **99**, 123902 (2007)
13. E. Oliva, Phys. Rev. A **84**, 013811 (2011)
14. J.P. Wittke, P.J. Warter, Pulse propagation in a laser amplifier. J. Appl. Phys. **33**, 1668 (1964)
15. O. Larroche, A. Klisnick, Phys. Rev. A **88**, 033815 (2013)

Chapter 9
EUV Laser Irradiation System with Intensity Monitor

M. Ishino, T.-H. Dinh, N. Hasegawa, K. Sakaue, T. Higashiguchi, S. Ichimaru, M. Hatayama, M. Washio, M. Nishikino and T. Kawachi

Abstract We have developed an extreme ultraviolet (EUV) laser irradiation system at QST. EUV laser has a wavelength of 13.9 nm, and this laser wavelength is close to the wavelength of EUV lithography ($\lambda = 13.5$ nm). The irradiation system has an intensity monitor based on the Mo/Si multilayer beam splitter. This intensity monitor provides the irradiation energy onto the sample surface. So it is possible to examine and confirm damage/ablation thresholds of optical elements and doses for the sensitivity of resist materials, which have the same specifications of those in the EUV lithography.

9.1 Introduction

The interaction between extreme ultraviolet (EUV) lights and materials, such as photoionization of gas [1], mass spectroscopy [2], and surface machining [3] are demonstrated in recent years. One of the attractive and important subjects in the laser–matter interaction is EUV laser ablation [4], because of their low damage threshold. Low ablation threshold results in the damage formation on the material surface easily, but leads to an efficient surface machining. In addition, the wavelength of EUV is short enough to draw nanometer scale patterns. And more, experimental results will give important information for the construction of ablation models. So, it is important to study the ablation phenomena of short-pulse lasers not only to

M. Ishino (✉) · T.-H. Dinh · N. Hasegawa · M. Nishikino · T. Kawachi
National Institutes for Quantum and Radiological Science and Technology, Kizugawa, Kyoto, Japan
e-mail: ishino.masahiko@gst.go.jp

K. Sakaue · M. Washio
Waseda University, Shinjuku, Tokyo, Japan

T. Higashiguchi
Utsunomiya University, Utsunomiya, Tochigi, Japan

S. Ichimaru · M. Hatayama
NTT Advanced Technology, Atsugi, Kanagawa, Japan

M. Kozlová and J. Nejdl (eds.), *X-Ray Lasers 2018*, Springer Proceedings in Physics 241, https://doi.org/10.1007/978-3-030-35453-4_9

describe the ablation dynamics but also to make nanometer scale drawing with three-dimensional structures.

To understand the EUV ablation deeply, we must correct the accurate data. Then, the ablation experiment must be carried out by using a reliable apparatus. The important parameter in the ablation experiment is the irradiation fluence, which directly reflects the character of the target material. To discuss the fluence, we must measure the spot size and intensity of the EUV laser pulse on the target. In this paper, we report the development of the EUV laser irradiation system with the intensity monitor at the National Institutes for Quantum and Radiological Science and Technology (QST).

9.2 Instrumentation

9.2.1 *EUV Laser Irradiation System*

The irradiation system has been installed in the soft X-ray laser facility at QST. This system was composed of five parts as follows; the EUV laser source, the attenuator composed by Zr filters, the intensity monitor using the Mo/Si multilayer beam splitter, the Mo/Si multilayer-coated spherical mirror, and the sample stage. Figure 9.1 shows the schematic diagram of the irradiation system.

The spatially coherent EUV laser pulse was generated from Ag plasma mediums. Ag tape targets were irradiated by the linearly focused infrared laser pulses generated from a chirped pulse amplification laser system with slab Nd: glass amplifiers. The

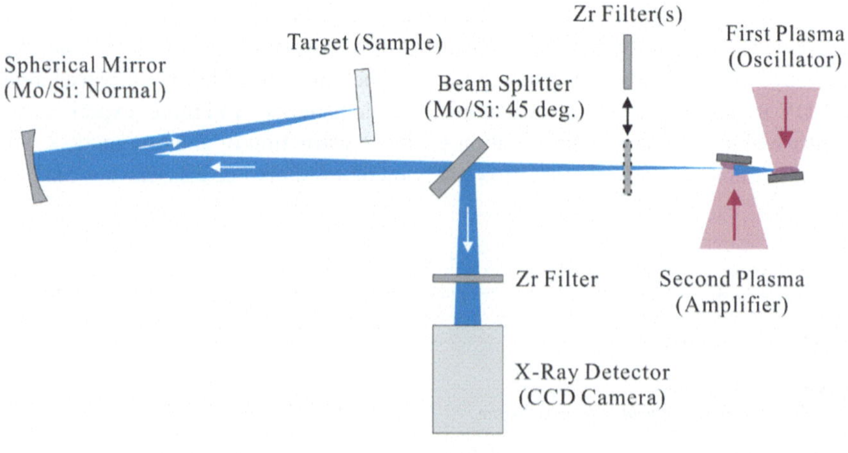

Fig. 9.1 Schematic diagram of the EUV laser irradiation system. The system has five components; EUV laser source, Zr filter attenuator, intensity monitor, focusing mirror, and sample stage

EUV laser pulse had a wavelength of 13.9 nm, pulse duration of 7 ps, and output energy of around 200 nJ/pulse. This laser system worked in a 0.1 Hz regime.

The intensity of the EUV laser pulse was reduced by Zr filters having various thicknesses. Zr filters were able to insert alone or in combination at the next of the EUV laser source. Zr filters reduced laser intensity, and could cut the almost scattered optical radiation from Ag plasma mediums.

The intensity monitor was composed by the Mo/Si multilayer beam splitter, Zr filters, and the X-ray CCD camera. The beam splitter was designed for an EUV laser of 13.9 nm at the incident angle of 45°. The EUV laser pulse was divided into two beams; one was the transmitted component through the splitter, which was toward the spherical mirror, and another one was reflected component, which was toward the X-ray CCD camera. The CCD camera worked as the intensity monitor. Zr filters installed between the beam splitter and the CCD camera adjusted the beam intensity so that the counter value of the CCD camera was not to be saturated. Both of transmittances of Zr filters and the sensitivity of CCD camera were evaluated by use of the EUV beamline at the synchrotron radiation facility of KEK-PF, Japan [5].

The EUV laser pulse toward the spherical mirror was reflected and focused onto the target surface. The spherical mirror had a radius of 1000 mm in curvature and placed at a distance of about 2,600 mm from the EUV laser output. The spherical mirror surface was coated by the Mo/Si multilayer. The Mo/Si multilayer was optimized for soft X-rays of 13.9 nm at a normal incidence angle.

The target material was set on the sample holder and the sample holder was mounted on the sample stage. The sample stage had two movable directions. To confirm the focal patterns and the best focal position, the sample target was moved along the propagation direction of the EUV laser beam. To irradiate the fresh surface, the stage was moved perpendicular to the propagation direction, after the prescribed number of shots.

9.2.2 Intensity Calibration

To decide the irradiation energy onto the target surface from the acquired energy by the intensity monitor, the relationship between the irradiation energy and the monitored energy had to be evaluated. Before the evaluation, the polarization of EUV laser, i.e., the vertical and the horizontal components, which were in accordance with the *s*- and *p*- polarizations with respect to the Mo/Si multilayer beam splitter, had to be considered. Because the Mo/Si multilayer mirror worked as the polarizer due to the incident angle of 45°., which was near the pseudo-Brewster angle, the reflected EUV laser toward the CCD camera included the vertical polarization only. So that the intensity monitor measured the vertical component of the polarization of the EUV laser always. However, the beam splitter transmitted the horizontal components and the remaining part of the vertical component of the EUV laser, excepting of absorbed intensity by the beam splitter.

To measure the irradiation energy, the Mo/Si multilayer-coated plane mirror and the X-ray CCD camera were settled after the spherical mirror (just before the sample stage). The Mo/Si multilayer on the plane mirror was optimized for soft X-rays of 13.9 nm at the incidence angle of 45 deg., so that the Mo/Si multilayer could reflect the *s*-polarization with respect to the plane mirror. When the plane mirror set to reflect the vertical component, the relative intensity of the vertical (irradiation)—vertical (monitor) components could be compared. If the plane mirror is set to reflect the horizontal component, the relative intensity of the horizontal (irradiation)—vertical (monitor) components could be measured. The reflectivity of the Mo/Si multilayer plane mirror and the sensitivity of the CCD camera used in this setup were also evaluated described above.

Figure 9.2 shows the relative intensities between the intensity of the sample irradiation vertical (left) and the horizontal (right) components with respect to the monitored intensity, respectively. The vertical component of the EUV laser energy irradiated onto the target surface was about 0.30 times smaller than the monitored energy. The horizontal component of the irradiation energy was as small as 0.42 times of the monitored energy. So the irradiation energy onto the sample surface was evaluated to be (*Irradiation Energy*) = (0.30 + 0.42) x (*Monitor Energy*). The EUV laser irradiation experiments were carried out by the use of this evaluated value.

From the irradiation energy measurement obtained above, we have noticed that the intensity dispersion of the horizontal component to the CCD value is larger than that of the vertical component. This fact means that the polarization of the EUV laser is unstable, and the EUV laser has no stabilized polarization direction. We will discuss the polarization of the EUV laser elsewhere.

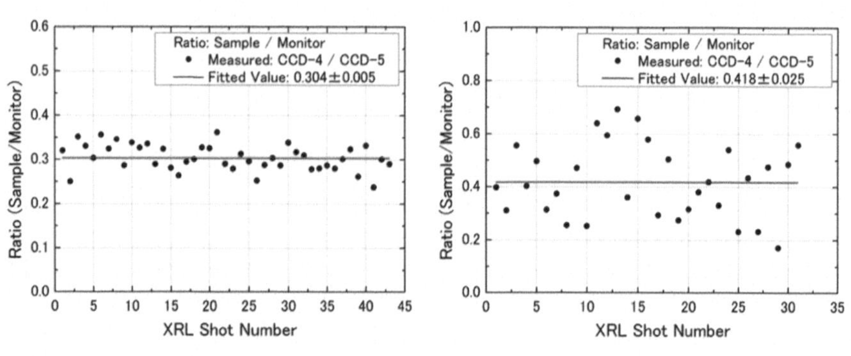

Fig. 9.2 Relative intensity of the vertical component (left) and the horizontal component (right) with respect to the monitor value

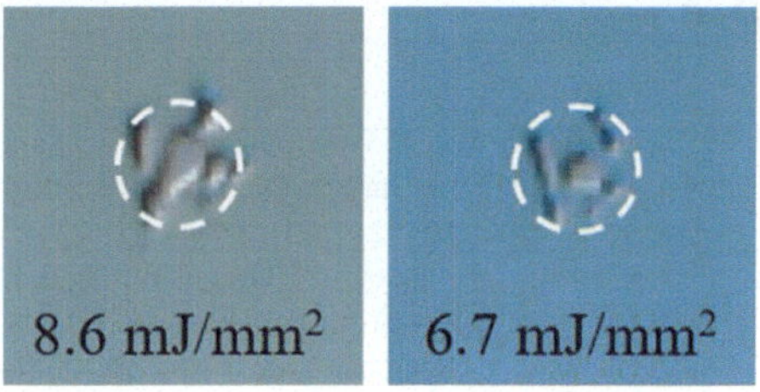

Fig. 9.3 Damage structures on PMMA film. White circles show the average focal spots of about 10 μm in diameter

9.3 EUV Laser Irradiation Experiments

We had the irradiation experiments by the use of the irradiation system. To acquire the EUV laser images by use of the X-ray CCD camera on the intensity monitor, we could evaluate the irradiation energy onto the target surface.

Figure 9.3 demonstrates two images of damage structures recorded in the PMMA film. These damage structures were formed by EUV laser pulses with average fluences of 8.6 and 6.7 mJ/cm^2. The average fluence was derived from the irradiation energy divided by the average focal spot size ($\Phi \sim 10$ μm), so the evaluated values described above were a little bit small (underestimation). The confirmation of the correct position, also the exact diameter, of the best focal spot were next plane, at this time.

9.4 Summary

We developed the EUV laser irradiation beamline. The irradiation system has an intensity monitor based on the Mo/Si multilayer beam splitter. This intensity monitor provides the irradiation energy onto the sample surface. In addition, we have the plan to install the Ru mirror-based EUV attenuator to reduce the laser intensity continuously [6]. The EUV attenuator consists of a pair of Ru mirrors. By using the EUV attenuator, we can reduce the EUV laser intensity continuously and sensitivity.

The EUV laser has a wavelength of 13.9 nm, and this laser wavelength is close to the wavelength of the EUV lithography of 13.5 nm. So it is possible to examine and confirm damage/ablation thresholds of optical elements and doses for the sensitivity of resist materials, which have the same specifications of those in the EUV lithography [7].

Acknowledgements The authors acknowledge the financial support from the research program of "QST Mirai-Lab". We are also grateful to H. Hara and H. Kawasaki of Utsunomiya Univ., Y. Yamamoto and T. Sasaki of QST Kansai for their support and Y. Hosaka, K. Yoshimura and H. Yamamoto of QST Takasaki for collaboration in this work.

References

1. I. Sabar, A. Bartnik, P. Wachulak, W. Skrzeczanowski, R. Jarocki, H. Fiedorowicz, Photoionization of atomic Neon induced using nanosecond phlses of extreme ultraviolet (EUV). Springer Proc. Phys. **202**, 203–211 (2018)
2. C.S. Menoni, I. Kuznetsov, T. Green, W. Chao, E.R. Bernstein, D.C. Crick, J.J. Rocca, Soft x-ray laser ablation mass spectroscopy for chemical composition imaging in three dimensions (3D) at the nanoscale. Springer Proc. Phys. **202**, 221–230 (2018)
3. H. Urai, T. Makimura, M. Ogawa, Micrometer-scale photo-direct machining of polydimethulsiloxane using laser plasma EUV radiations. Springer Proc. Phys. **202**, 383–386 (2018)
4. K. Mikami, M. Ishino, T.-H. Dinh, Y. Yamamoto, N. Hasegawa, M. Nishikino, T. Kawachi, S. Motokoshi, T. Jitsuno, Laser-induced damage on silica glasses by irradiation of soft x-ray laser pulse. Springer Proc. Phys. **202**, 333–337 (2018)
5. T.-H. Dinh, Y. Kondo, T. Tamura, Y. Ono, H. Hara, H. Oikawa, Y. Yamamoto, M. Ishino, M. Nishikino, T. Makimura, P. Dunne, G. O'Sullivan, S. Ohta, K. Kitano, T. Ejima, T. Hatano, T. Higashiguchi, Evaluation of a flat-field grazing incidence spectrometer for highly charged ion plasma emission in soft x-ray spectral region from 1 to 10 nm. Rev. Sci. Instrum. **87**, 123106 (2016)
6. M. Hatayama, S. Ichimaru, M. Ishino, M. Nishikino, S. Oku, Ru mirror based EUV attenuator with continuous and wide dynamic range. in Abstract PD1-06, International Conference on Synchrotron Radiation Instrumentation (2018)
7. S. Ichimaru, M. Ishino, M. Nishikino, M. Hatayama, N. Hasegawa, T. Kawachi, T. Maruyama, K. Inokuma, M. Zenba, S. Oku, Irradiation damage test of Mo/Si, Ru/Si, and Nb/Si multilayers using the soft x-ray laser built at QST. Springer Proc. Phys. **202**, 303–308 (2018)

Chapter 10
Li-like Aluminum Plasma Soft X-ray Laser by Means of Recombination Scheme

S. Namba, N. Kubo, C. John, N. Hasegawa, M. Kishimoto, M. Nishikino and T. Kawachi

Abstract In order to realize a compact and high repetition rate plasma X-ray laser, we have developed a recombination scheme plasma X-ray laser. In particular, we focused on a laser oscillation of Li-like aluminum $3d$-$4f$ transition (15.5 nm), which was already demonstrated in a high-density and low-temperature Al laser plasma with a low gain coefficient. In this study, by using an Nd:YAG laser (500 ps, 3 J, 16 pulse-trains) we created the lasing media and succeeded in laser oscillation at 15.5 nm. The optimization of the pulse-train in terms of the pulse duration and its interval enhanced the gain coefficient to $g = 4.3\ \mathrm{cm}^{-1}$.

10.1 Introduction

X-ray laser is applied to many scientific and engineering fields such as X-ray microscope, X-ray photoelectron spectroscopy, X-ray lithography, and X-ray holography [1]. Because the plasma-excited X-ray laser is driven by the stimulated emission phenomenon of multiply charged ions in laser-generated plasmas, a driver laser system with extremely high output energy is necessary [2]. Therefore, the development of a compact and high repetition X-ray laser, which can be operated and maintained even in a university laboratory, has been carried out so far [3, 4]. As major lasing schemes for the X-ray laser, there are two methods for the creation of a population inversion, that is, a transient collisional excitation (TCE) scheme and a recombination plasma scheme.

In the recombination plasma scheme that we have focused on, a relatively light element is ionized by laser irradiation onto metal or gas target, and a high-temperature and high-density plasma are generated. When the hot-dense plasma is rapidly cooled

S. Namba (✉) · N. Kubo · C. John
Graduate School of Engineering, Hiroshima University, 1-4-1 Kagamiyama, Higashihiroshima, Hiroshima, Japan
e-mail: namba@hiroshima-u.ac.jp

N. Hasegawa · M. Kishimoto · M. Nishikino · T. Kawachi
Kansai Photon Science Institute, National Institute for Quantum and Radiological Science and Technology, 8-1 Umemidai, Kizu, Kyoto 619-0215, Japan

M. Kozlová and J. Nejdl (eds.), *X-Ray Lasers 2018*, Springer Proceedings in Physics 241, https://doi.org/10.1007/978-3-030-35453-4_10

due to an adiabatic expansion, a nonequilibrium plasma is created, where a three-body recombination process dominates over the other reactions. Subsequently, the electrons captured into highly excited states (Rydberg states) are subjected to collisional de-excitation, resulting in the population flow into the lower states. Consequently, a population inversion between certain lower levels is generated. The first demonstration of the recombination plasma X-ray laser was made by Suckewer et al. where they succeeded in lasing H-like C $n = 3 \rightarrow 2$ transitions (18.2 nm) by using a kJ-class CO_2 driver laser [5].

On the other hand, Hara et al. has also adopted the recombination plasma scheme and observed the soft X-ray amplification of Li-like Al $3d$-$4f$ transition (15.5 nm) [6–8]. The method to create a lasing plasma medium, a joule class compact Nd:YAG laser system was used, in which 16 pulse-trains having each pulse duration of 100 ps and their interval of 200 ps generated the Al laser plasma. This pulse train method can produce the highly charged ion up to He-like Al ion efficiently. Recently, they demonstrated that the gain coefficient measured for the 15.47 nm line was 3.2 cm^{-1} and the duration of the gain was longer than 1 ns. However, the optimization of the pulse width and its interval was not made. In this study, therefore, a recombination X-ray laser by 16 laser pulses of a 500-ps width and an interval of 1 ns was examined to enhance the gain coefficient of the soft X-ray laser.

10.2 Experimental Setup

Figure 10.1 shows a schematic diagram of the soft X-ray laser experiment A mode-locked 1064-nm YAG oscillator having a pulse duration of 500 ps was divided into 16-pulse-train by an optical stacker and a delay line component. By employing the pulse-train laser irradiation, highly charged ions are produced efficiently through continuous electron heating, and that the plasma temperature drops significantly immediately after the laser irradiation. Thus, the pulse-train laser provides a favorable plasma condition to obtain a high-gain coefficient owing to the three-body collisional recombination. The interval of each pulse was ~1 ns, which was adjusted by an auto-correlator. Then, the laser pulse was delivered into the different two amplifiers,

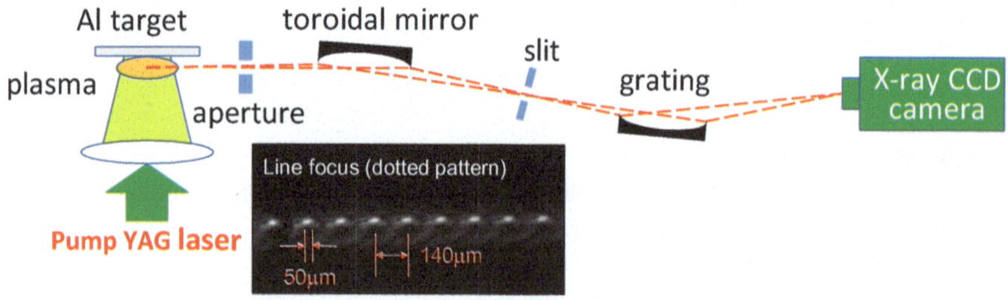

Fig. 10.1 Schematic of the experimental setup. The line focusing pattern with intense microdots is also shown

resulting in the former eight pulses were more than four times higher than the latter intensity. The total pulse energy of up to 3 J was irradiated onto the Al target. For a line focusing, a prism lens array was used [9]. The focusing assembly consists of a segmented prism, a beam expander and a cylindrical lens, and forms a 10-mm-line focus on the target. Figure 10.1 shows that the irradiation pattern had small dots due to an interference effect of the laser beam from each segment, and was about 50 μm diameter with a spacing of 140 μm, yielding the power density of ~2×10^{12} W/cm^2.

The soft X-ray spectra collected by a toroidal mirror were measured by using flat-field grazing (1200 grooves/mm). The spatial resolution was less than 150 μm in the direction normal to the target surface while for the perpendicular direction it was ~50 μm in the direction perpendicular to the target normal. The detector was a back-illuminated cooled X-ray CCD camera. The wavelength calibration of the spectrometer was performed by measurement of line spectra attributed to carbon spectra.

10.3 Results and Discussion

Figure 10.2 shows the soft X-ray CCD image (a) and its spectrum (b) around 15 nm emitted from Al laser plasma. In this measurement, the total laser energy was ~2 J and the laser irradiation length was 10 mm with a flat intensity distribution owing to the prism lens array. The entrance slit was set to be 30 μm. With increasing irradiation length, the emission intensity of 15.5 nm attributed to Li-like Al ion 3*d*-4*f* transition became brighter, indicating that the lasing action of this transition seems to occur, as will be described later in detail.

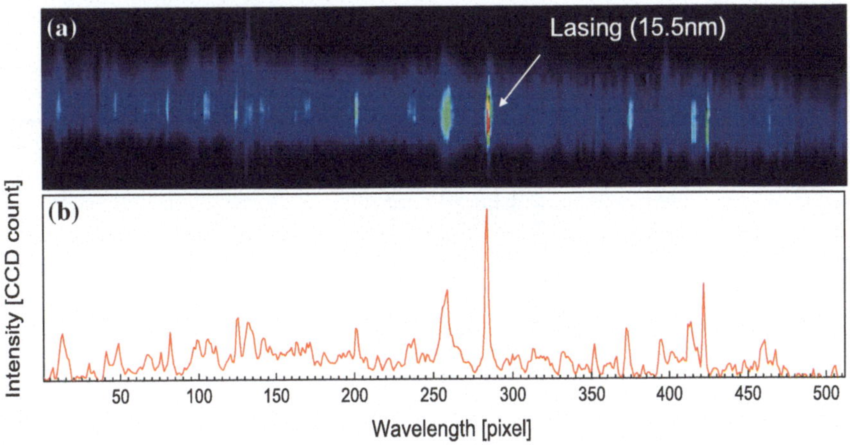

Fig. 10.2 **a** Soft X-ray CCD image and **b** corresponding spectrum observed at 2-J laser irradiation

Figure 10.3 plots the spatial distribution of the 15.5-nm X-ray yield in the direction normal to the target surface. The intense radiation of a 15.5-nm lasing line is observed in a spatial region of 100–650 μm from the target surface and the highest emission intensity is obtained around 150 μm, around which the optimal plasma condition for the lasing is probably fulfilled. This result also suggests that the initial plasma at the target surface should have too high-density and high-temperature to create the rapid cooling recombination plasma, leading to a formation of population inversion between 3*d*-4*f* states.

In order to verify the laser oscillation and evaluate the gain coefficient, we measured Soft X-ray yield for the irradiation length of 4.0, 7.5, and 10.0 mm, as shown in Fig. 10.4. The observations were made at 300 μm from the target surface and total laser energy of 2 J. From the fitting curve, the gain coefficient is evaluated to be $g = 4.3 \pm 0.4\ \mathrm{cm}^{-1}$. Considering that the gain obtained in the previous research was $g = 3.2\ \mathrm{cm}^{-1}$, the present pulse train can create high-density and low-temperature recombining plasma, resulting in large population inversion.

10.4 Summary

In order to enhance the gain coefficient of Li-like Al X-ray laser (15.5 nm), we have changed the property of the 16 pulse-train, especially, with a pulse duration of 500 ps and its interval of 1 ns. The Al plasma generated with the laser pulses seemed to be a suitable condition for creating the population inversion due to the three-body collisional recombination process. Consequently, the gain coefficient of $g = 4.3\ \mathrm{cm}^{-1}$ was achieved, which was higher than that obtained by the previous experiment with a 100-ps pulse width and 200-ps interval ($g = 3.2\ \mathrm{cm}^{-1}$).

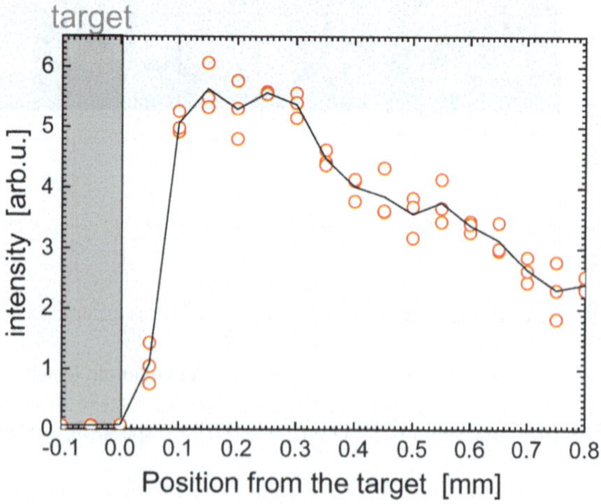

Fig. 10.3 Spatial distribution of the 15.5-nm emission intensity

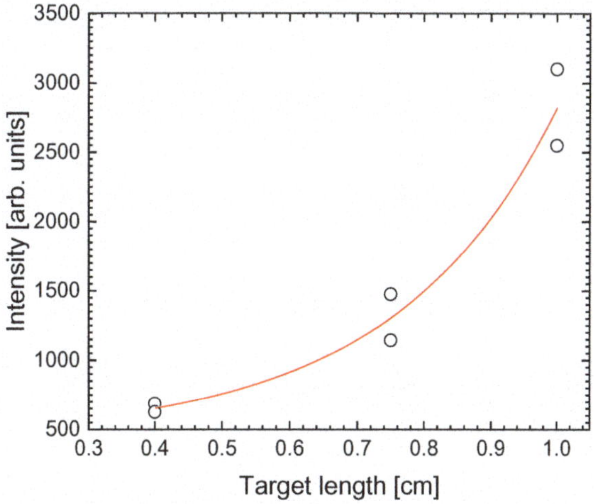

Fig. 10.4 X-ray laser intensity against the target length measured at 300 μm from the target surface. The gain coefficient is evaluated to be $g \sim 4.3$

Acknowledgements This work was supported by JKA and its promotion funds from AUTO RACE.

References

1. D. Attwood, *Soft-X-rays and Extreme Ultraviolet Radiation: Principles and Applications* (Cambridge University Press, 2000)
2. R.C. Elton, *X-ray Lasers* (Academic Press, London, 1990)
3. H. Daido, Rep. Prog. Phys. **65**, 1513 (2002)
4. S. Suckewer, P. Jaegle, X-Ray laser: past, present, and future. Laser Phys. Lett. **6**, 411 (2009)
5. S. Suckewer, C.H. Skinner, H. Milchberg, C. Keane, D. Voorhees, Phys. Rev. Lett. **55**, 1753 (1985)
6. T. Hara, K. Ando et al., Jpn. J. Appl. Phys. **28**, L1010 (1993)
7. H. Hirose, T. Hara, K. Ando et al., Jpn. J. Appl. Phys. **32**, L1538 (1993)
8. N. Yamaguchi, T. Hara et al., Jpn. J. Appl. Phys. **38**, 5114 (1999)
9. N. Yamaguchi, C. Fujikawa, T. Ohchi, T. Hara, Jpn. J. Appl. Phys. **39**, 5268 (2000)

Chapter 11
Experimental Study of the Gain Dependence on Driving Laser Pulse Leading Edge Duration

P. Zakopal, M. Krůs and M. Kozlová

Abstract We present the preliminary results of the experiment focused on the generation of the Ne-like Ti XUV laser radiation (32.6 nm). Within the experiment, the two-pulse scheme (pre-pulse and main pulse) was adopted. The population inversion is produced between $2p^5 3p$ and $2p^5 3$ s fine-structure levels of Ne-like ions, through electron collisional excitation from the Ne-like ground state $2p^6$. The inversion is maintained by strong radiative dipole decay from the 3 s levels back to the ground state while the upper 3p levels are metastable with respect to this process. Under appropriate plasma conditions, the strongest population inversion is generated (due to the highest monopole electron excitation rate from the ground state) for $(2p^5_{1/2} 3p_{1/2})$ $(J = 0) \rightarrow (2p^5_{1/2} 3s_{1/2})$ $(J = 1)$ transition in Ne-like Ti, corresponding to the wavelength of 32.62 nm. The intensity and beam profile emitted by this type of XRL strongly depends on the density profile of the amplifying plasma, as the largest population inversion on the $J = 0 \rightarrow 1$ transition occurs at high electron densities. Hence various laser beam parameters (i.e. intensity ratio, delay between pulses, etc.) were investigated. The main focus was to study the dependence between the XRL and the pulse duration of the driving laser beam, especially the leading edge of the pulse.

11.1 Introduction

In recent years, soft x-ray lasers have established themselves as a novel tool for applications such as protein and living cell diffractometry [1], dense plasma interferometry [2] or probing ultrafast processes using high-harmonic generation scheme [3]. While many approaches to generate coherent XUV radiation exist, from large free-electron laser (FEL) facilities with high-brightness sources [4] over plasma-based laser-driven lasers [5] to capillary discharge lasers operable as table-top sources [6], the common

P. Zakopal (✉) · M. Krůs · M. Kozlová
Institute of Plasma Physics CAS, Za Slovankou 1782/3, Prague 182 00, Czech Republic
e-mail: zakopal@ipp.cas.cz

M. Kozlová
Institute of Physics CAS, Na Slovance 1999/2, Prague 182 21, Czech Republic

M. Kozlová and J. Nejdl (eds.), *X-Ray Lasers 2018*, Springer Proceedings in Physics 241, https://doi.org/10.1007/978-3-030-35453-4_11

goal of x-ray laser research is improvement in overall efficiency, brightness and both temporal and spatial coherence.

With more than decade-long experience in solid target x-ray laser generation [7], PALS Research Centre is well suited for studies of their pumping conditions optimization. In this paper, we discuss our preliminary experimental results to maximize the XRL photon yield by optimization of driving pulse leading edge duration.

11.2 Experimental Setup

A conventional 2-pulse scheme for quasi-steady state (QSS) XUV lasing from a solid target—a 30 mm long superpolished titanium slab—was used. Both pre-pulse and main pulse were produced by kJ-class iodine laser at PALS laboratory (typical pulse duration $\tau = 250$ ps, $P = 3$ TW @1315 nm). Laser beam was split using a 60:40 beamsplitter before passing through final amplifier, giving a maximal pre-pulse energy of 80 J and main pulse energy of 500 J. Pre-pulse was incident on target at 30° to normal, main pulse at target normal incidence, both on overlapping line foci produced by a combination of positive spherical lens and an array of cylindrical negative lenses. X-ray laser pulses radiating from resulting plasma were directed to flat-field XUV spectrometer, filtered by 0.2 μm Al foil.

In order to obtain a spatial profile, the titanium slab was rotated within 1° of the original target, and normal and resultant intensity profile is plotted in Fig. 11.1. In order to obtain the laser gain, the target effective length was varied between 15 and 30 mm. Main driving pulse and pre-pulse mutual delay were varied between 5 and 15 ns to study effects on Ne-like XUV spectral line intensity (I_{XUV}), for each target length and rotation as fixed parameters. The influence of shot-to-shot variation of main pulse duration, especially leading edge duration, was analyzed simultaneously using infrared streak camera.

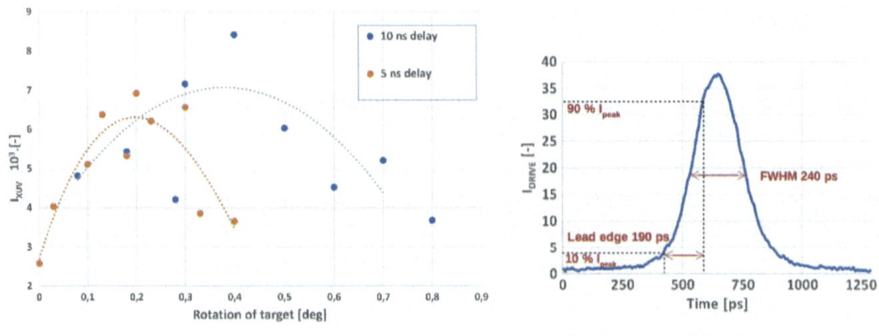

Fig. 11.1 (left) Ne-like Ti spectral line intensity I_{XUV} as a function of target slab rotation, for pre-pulse–main pulse delay of 5 and 10 ns, respectively, with gaussian fits. (right) Typical main driving laser pulse temporal profile and characteristics

11.3 Pumping Optimization Measurements

Typical driving pulse temporal profile from the infrared streak camera (Fig. 11.1 right) shows a Gaussian pulse shape with FWHM of ~240 ps and leading edge duration ~190 ps (defined here as rise time from 10 to 90% of peak intensity). Figure 11.1 (left) shows line intensity I_{XUV} varying with target slab rotation during a scan of XUV beam axis. In the small angular range used here (within 1°), the target slab rotation approximates the effect of rotating and shifting the XUV spectrometer with respect to target as in a virtual detector array. Of interest is the higher angular radius of the XUV beam with longer pre-pulse–main pulse delay, due to the radiating plasma having more time to spread away from target line focus and producing a more broad, divergent beam.

After the intensity scan revealed true XUV beam axis, measurement of spectral line intensity I_{XUV} as a function of different pumping parameters was carried out. Figure 11.2 shows typical spectrum of XUV pulse filtered by 0.2 μm Al foil, with dominant 32.6 nm line resulting from $(2p^5_{1/2}3p_{1/2})$ $(J = 0) \rightarrow (2p^5_{1/2}3s_{1/2})$ $(J = 1)$ transition in Ne-like Ti.

For constant target rotation at XUV axis and for pre-pulse–main pulse delay of 10 ns, the integrated Ne-like line intensity was analyzed with respect to driving pulse

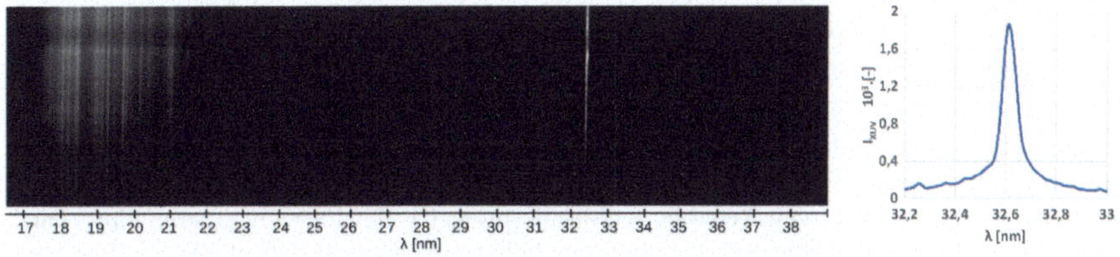

Fig. 11.2 (left) Spectrum from XUV flat-field spectrometer. (right) 32.6 nm Ne-like line intensity profile

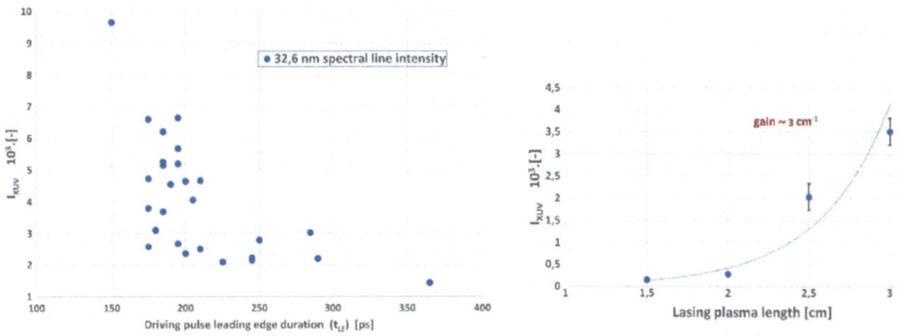

Fig. 11.3 (left) Intensity of Ne-like Ti laser line as a function of driving pulse leading edge duration (right). Output XRL line intensity as a function of lasing plasma length with lasing gain fit

leading edge duration. Results (Fig. 11.3 left) show an upward trend in I_{XUV} at shorter leading edge.

Moreover, we changed the effective target length from 15 to 30 mm by shifting the target slab away from laser line foci in order to investigate XUV pulse generation in differing conditions and calculate the laser gain. The gain of ~3 cm^{-1} obtained by fitting the intensity/plasma length (Fig. 11.3 right) is a rough estimate, but shows the XRL is below saturation, with saturation in Ne-like plasmas usually occurring at *gl* ~16 [5].

11.4 Conclusion

Successful generation of XUV Ne-like laser pulses using pre-pulse–main pulse, quasi-steady state pumping scheme was carried out at PALS kJ-class iodine laser. Observed I_{XUV} dependence on main pulse leading edge duration shows higher XUV gain for shorter drive pulses. Similarly, shorter overall driving pulse duration leads to higher XUV radiation due to temporal symmetry of used driving pulse. The ability to properly tune leading edge duration of driving pulse hence shows to be important in an effort to maximize XUV laser outputs in QSS scheme. As our analysis was performed only on shot-to-shot pulse length variations, future experiments where active control over driving pulse duration and/or temporal profile would provide more information on optimizing the XUV output.

Acknowledgements Authors wish to acknowledge financial support from Czech Science Foundation (GAČR) project 18-27340S. The financial support provided by the Ministry of Education, Youth and Sports of the Czech Republic within the project LM2015083 is greatly appreciated. The experiment was carried out with the important help from technical staff of PALS iodine laser. Support from PALS Research Institute is also acknowledged.

References

1. M. Seibert et al., Single mimivirus particles intercepted and imaged with an X-ray laser. Nature **470**, 78–81 (2011)
2. L.M.R. Gartside et al., Extreme ultraviolet interferometry of warm dense matter in laser plasmas. Opt. Lett. **35**, 3820–3822 (2010)
3. T. Popmintchev, M.-C. Chen, P. Arpin, M.M. Murnane, H.C. Kapteyn, The attosecond nonlinear optics of bright coherent X-ray generation. Nat. Photonics **4**, 822–832 (2010)
4. B.W.J. McNeil, N.R. Thompson, X-ray free-electron lasers. Nat. Photonics **4**, 814–821 (2010)
5. P. Jaeglé, Coherent sources of XUV radiation: soft X-Ray lasers and high-order harmonic generation (Springer, 2007)
6. J.J. Rocca, V. Shlyaptsev, F.G. Tomasel, O.D. Cortázar, D. Hartshorn, J.L.A. Chilla, Demonstration of a discharge pumped table-top soft-X-ray laser. Phys. Rev. Lett. **73**, 2192 (1994)
7. B. Rus, T. Mocek, A.R. Präg, M. Kozlová, G. Jamelot, A. Carillon, D. Ros, D. Joyeux, D. Phalippou, Multimillijoule, highly coherent x-ray laser at 21 nm operating in deep saturation through double-pass amplification. Phys. Rev. A **66**, 063806 (2002)

Part II
High-Order Harmonic Generation

Chapter 12
High-Charge XUV Vortex Beam: Generation and Hartmann Wavefront Sensor Characterization

F. Sanson, A. K. Pandey, F. Harms, G. Dovillaire, E. Baynard, J. Demailly, O. Guilbaud, B. Lucas, O. Neveu, E. Papagionnouli, M. Pittman, D. Ros, M. Richardson, E. Johnson, W. Li, Ph. Balcou and S. Kazamias

Abstract In this paper, we present the characterization of high charge vortex structure of an OAM beam produced through high-harmonic generation in a rare gas using an XUV Hartmann wavefront sensor. We show that the phase-matched absorption limited high-harmonic generation is able to retain the high-charge vortex structure of the XUV beam even in a rather long (1 cm) generation medium. Additionally, our recent results on the influence of infrared driver wavefront quality on the characteristics of XUV vortex beam are presented.

F. Sanson (✉) · A. K. Pandey · J. Demailly · O. Guilbaud · B. Lucas · O. Neveu · E. Papagionnouli · D. Ros · S. Kazamias
Laboratoire de Physique des Gaz et des Plasmas, Université Paris-Saclay (UMR-CNRS 8578), Bat 210, Rue Becquerel, F-91405 Orsay, France
e-mail: fabrice.sanson@u-psud.fr

F. Sanson
Amplitude Laser Group, Scientific Business Unit-Lisses Operations, 2/4 Rue du Bois Chaland, 91090 Lisses, France

F. Harms · G. Dovillaire
Imagine Optic, 18, rue Charles de Gaulle, 91400 Orsay, France

E. Baynard · M. Pittman
LASERIX, Centre Laser Université Paris Sud, FR2764, Université Paris-Saclay (UMR-CNRS 8578), Bat 106, 91405 Orsay, France

M. Richardson
University of Central Florida, 4304 Scorpius Street, Orlando, FL 32816-2700, USA

M. Richardson · Ph. Balcou
Centre Laser Intense et Applications, Université de Bordeaux (UMR-CNRS-CEA 5107), Talence, France

E. Johnson · W. Li
Department of Electrical and Computer Engineering, Clemson University, Clemson, SC 29634, USA

M. Kozlová and J. Nejdl (eds.), *X-Ray Lasers 2018*, Springer Proceedings in Physics 241, https://doi.org/10.1007/978-3-030-35453-4_12

12.1 Introduction

Optical beams carrying a l Orbital Angular Momentum (OAM), i.e., with spatial phase $e^{il\theta}$, θ being the azimuthal angle, are currently a very active field of research due to their numerous prospective applications. The possibility to produce such kind of beam, especially in the Extreme UV domain, which has been demonstrated experimentally and studied theoretically [1], gives access to a completely new physics, particularly in the light-matter interaction field.

Recently we demonstrated for the first time the ability of an extreme ultraviolet (XUV) Hartmann wavefront device to measure a high-charge vortex beam created by high-harmonic generation. We have also shown that High-Harmonic Generation (HHG) in a rather long medium (1 cm) is able to maintain this high-charge vortex [2]. In the present work, we report a further detailed quantitative study of the influence of aberrations in the infrared driver beam on the orbital angular momentum transfer.

All those experiments were performed on the LASERIX platform from Université Paris-Saclay (France). This facility is historically dedicated to soft X-Ray laser developments but offers variety of different beamlines including the one we are using for HHG. This beamline is composed by a Ti:Sa laser chain at 10 Hz repetition rate emitting around 815 nm (center wavelength is managed by a couple of a dazzler and a mazzler devices) amplified until 60 mJ after first stage. Beam is then splitted, and HH beamline uses 25 mJ compressed separately to 40 fs with an $1/e^2$ diameter of 20 mm. We then guide this beam in our experimental setup (Fig. 12.1).

12.2 First Experiments

In our first experiment, the beam described previously was guided onto a $l = 1$ azimuthal charge 16 angular sector high fluence damage threshold phase plate. It is provided by Clemson University for the University of Central Florida. We then guided it through an iris and focused by a 1 m focal length lens into a 27 mbar argon gas cell, generating High-Harmonic beam. This beam is guided onto a Hartmann XUV sensor (2 m away from source) from the Imagine Optic company (France) using a multilayer dielectric mirror at 45° [4] prepared by the Institut d'Optique in Palaiseau (France) and designed to reflect $\lambda = 32.6$ nm corresponding to the 25th harmonic order (Fig. 12.1). This is done in order to ensure the analysis of a quasi-monochromatic wavefront.

In this configuration, we obtain the results shown in Fig. 12.2.

We found that HH beam generated by a quasi gaussian infrared driver is a pretty gaussian XUV beam carrying a small value of 0° astigmatism (around 0.04λ Peak to Valley) which does not visibly impact the intensity profile even with 2 m propagation.

Analyzing the wavefront, we found the l_q OAM order seems to follow $l_q = ql$, where q is the harmonic order and l the OAM order of the driver beam following [3]. Indeed when we subtract a $l = 25$ theoritical OAM to the experimental beam wave-

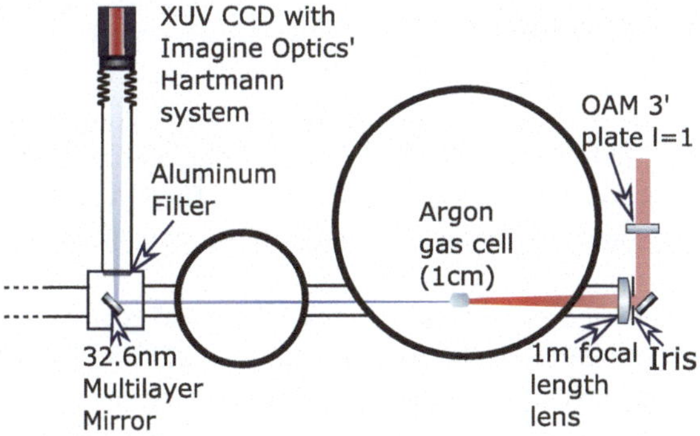

Fig. 12.1 First experimental setup

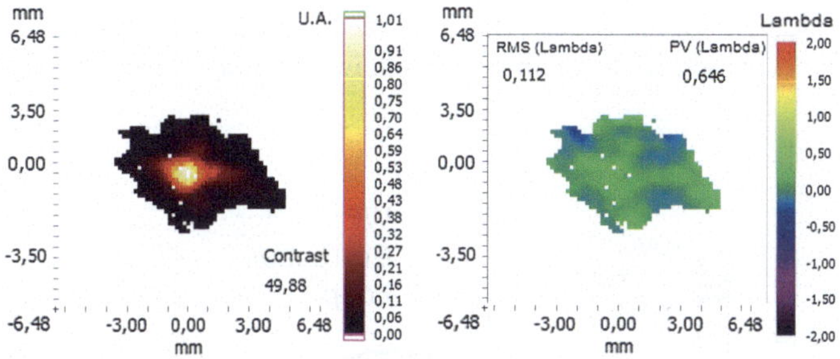

Fig. 12.2 Intensity (Left) and wavefront (Right) of classical high-harmonic generation. Acquisition time : 1 s, Iris : 11 mm, Energy before iris : 3.6 mJ. 1 λ corresponds to a phase shift of 2π. We give both Peak to Valley (PV) and RMS value of phase shift

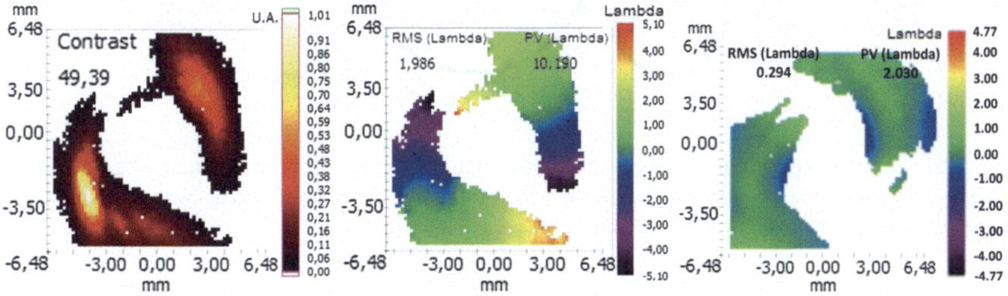

Fig. 12.3 Intensity (Left), wavefront (Center), and phase residual (Right) after substracting a l = 25 OAM. Acquisition time : 10 s, Iris : 10.5 mm, energy before iris 3.1 mJ

front we obtain the residual shown on the right side of Fig. 12.3, showing no more azimuthal variation of phase. Nevertheless Fig. 12.3 does not show the characteristic full donut beam but a bilobed profile, which we attribute to residual aberration in the infrared beam and which we demonstrated in the second campaign.

12.3 Second Experiment

For this experiment, we added a full corrective loop composed by an ILAO deformable mirror and an infrared HASO wavefront sensor (both from Imagine Optic). The ILAO is used close to normal incidence. The HASO is placed on a leak of a dielectric mirror and is imaging the plane of the beam at the position of the ILAO (Fig. 12.4).

The first step consisted in running a feedback loop and generating HH, obtaining the beam profiles presented in the first column of Fig. 12.5. The phase is clearly variying along azimuthal axis but the wavefront is still presenting a bilobed shape. By measuring the residual aberrations in the driver beam using a SID4 sensor (Phasics company (FRANCE)) we noticed that the value of astigmatism at 0° was around 150 nm, meaning that focusing lens or diagnostic beamline (by retrocontrol from the loop) was adding it on the beam. Afterward results presented in Fig. 12.5 show the beam profile evolution when adding deliberately a certain amount of astigmatism.

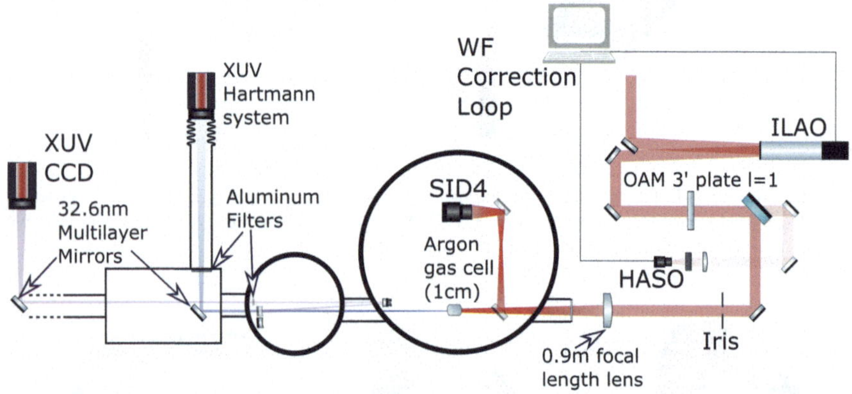

Fig. 12.4 Second experimental setup. Adding the diagnostic beamline (small energy beam sent on HASO) and corrective loop necessary for the use of the ILAO deformable optic. Another wavefront diagnostic (SID4) is used punctually to control wavefront correction

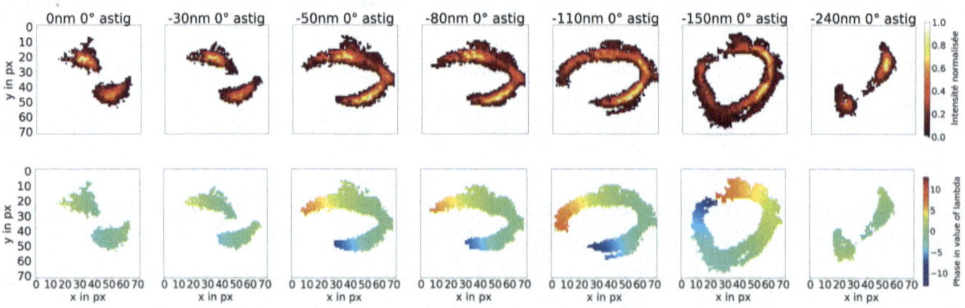

Fig. 12.5 Intensity (Top) and phase profile (Bottom) evolution with the value of 0° astigmatism in the driver IR pulse. Acquisition time : 0.5 s, Iris : 12 mm, Energy after iris 2.96 mJ

With the help of the deformable mirror, we succeeded in recovering a full ring for a value of astigmatism of –150 nm as compared to the initial value.

12.4 Study of the OAM

We then performed the same study as previously but were not able to recover the quantum law $l_q = ql$. In our case, we found average $l = 18$ by using a theoritical fit based on the wavefront and 18.2 using calculation based on the intensity profile as done in Géneaux et al. [3] (with huge error bar due to the method, ± 6.9).

At this point, we are suspecting different causes but we cannot affirm which one is explaining this unexpected value of l. According to the article by Rego et al. [5] this law $l_q = ql_1$ can be perturbated by the presence of a second OAM order l_2 in the driver beam, becoming $l_q = (p - n)l_1 + nl_2 + m(l_2 - l_1)$, with p and n the numbers of infrared photons from mode l_1 and l_2, respectively, which are absorbed to create the harmonic photon of energy $qh\nu_{IR}$, ($n + p = q = 25$ here) and m indicating the role of harmonic atomic phase. Such a presence could be an explanation here.

Indeed even after the correction by the deformable optic we still measure a residual of astigmatism in the harmonic beam. That means there is still residual astigmatism in the IR driver beam after correction (at level not easily correctable), yet this aberration can be interpretated as a creation of two OAM modes besides the central one, so here $l = -1$ and $l = 3$. Moreover, as we corrected as much as possible the aberration, we increased the maximum energy at the focal point and increase participation of the long path in the generation, playing on the m limit. These two elements could lead to this downscale of the real measured angular momentum.

12.5 Conclusion

We have shown that low values of 0° astigmatism with really small effects on a Gaussian beam generation can directly break the ring of an OAM beam and succeeded in recovering a full ring from an experimental beam using a deformable mirror. Nevertheless this new experiment showed us that it is not so easy to keep the high-order OAM during the HHG when playing with aberrations.

Acknowledgements The vortex optics used in this report were developed with the support of the U.S. Army Research Office. The project LIDEX OPT2X-Université Paris-Saclay funded the Hartmann XUV sensor. The PhD thesis of F. Sanson is funded through a CIFRE contract with the Amplitude Technologies company. The authors thank the Centrale d'Elaboration et de Métrologie des Optiques XUV (CEMOX) installation at Institut d'Optique Graduate School (IOGS) Palaiseau for the multilayer optics.

References

1. C. Hernández-García et al., Phys. Rev. Lett. **111**, 083602 (2013)
2. F. Sanson et al., Opt. Lett. **43**(12), 2780–2783 (2018)
3. R. Géneaux et al., Nat. Commun. **7**, 12583 (2016)
4. E. Meltchakov et al., Appl. Phys. A **98**, 111 (2009)
5. L. Rego et al., Phys. Rev. Lett. **117**, 163202 (2016)

Chapter 13
Ultrafast Inner-Shell Electron Excitation with High Energy Recollision Electron Driven by Mid-infrared Laser

Z. Zeng, Y. Deng, Y. Zheng, G. Marcus and R. Li

Abstract As one of the most important physical processes of strong-field laser-matter interaction, laser-driven electron-ion recollision is the fundamental process. As we have known, the well-known three-step model of HHG predicts that the cutoff law obeys $E_{\text{cutoff}} = I_{\text{p}} + 3.17U_{\text{p}}$, implying that the maximum kinetic energy of returning electron can be greatly extended by increasing the driving wavelength. With the long wavelength mid-infrared laser pulse, it is easy for the ponderomotive energy of the returning electron to be very large to excite the deep inner-shell electron, which may be used to investigate the ultrafast inner-shell electron dynamics.

13.1 Introduction

High-order harmonics generated by the interaction of extremely intense laser field with noble and simple polyatomic gases have been extensively studied and utilized to produce an intense coherent XUV or X-ray light source [1–3] to synthesize isolated attosecond pulses (IAPs) or attosecond pulse trains (APTs) by synchronizing harmonics near the cutoff region [4] to probe ultrafast dynamics of tunneling ionization (TI) and rescattering of electron wavepacket (EWP) of molecules and atoms with attosecond precision [5–8]. The process, which is frequently referred to as high-order harmonic generation (HHG), has been intuitively clarified by Corkum's three-step model [9]. The three-step model of HHG predicts that the cutoff law obeys $E_{\text{cutoff}} = I_{\text{p}} + 3.17U_{\text{p}}$ (I_{p} is the ionization potential. $U_{\text{p}} \propto I\lambda^2$ is the ponderomotive energy), implying that the maximum kinetic energy of returning electron can

Z. Zeng (✉) · Y. Zheng · R. Li
State Key Laboratory of High Field Laser Physics, Shanghai Institute of Optics and Fine Mechanics, Chinese Academy of Sciences, Shanghai 201800, China
e-mail: zhinan_zeng@mail.siom.ac.cn

Y. Deng
SwissFEL, Paul Scherrer Institut (PSI), 5232 Villigen, Switzerland

G. Marcus
Department of Applied Physics, Benin School of Engineering and Computer Science, Hebrew University of Jerusalem, 91904 Jerusalem, Israel

M. Kozlová and J. Nejdl (eds.), *X-Ray Lasers 2018*, Springer Proceedings in Physics 241, https://doi.org/10.1007/978-3-030-35453-4_13

be greatly extended by increasing the driving wavelength. With the long wavelength mid-infrared laser pulse, it is easy for the kinetic energy of the returning electron to be very large to excite the deep inner-shell electron.

The excitations of electrons from deeper shells are usually accompanied by multielectron dynamics such as double excitation, the Auger decay, Cooper minima, and the giant resonance, which cannot be explained by the single active electron approximation. Such excitations are unstable and usually decay on a time scale ranging from few femtoseconds to a few attoseconds [10, 11]. The decay may take place in a single step, but more often occurs as a cascade of radiative and nonradiative channels. Spectroscopic data may give some general information about the nature of such dynamics but often fail to follow the exact details, for example, the line widths of the cascade Auger decays reveal the total decay rate but not the order of decaying channels and their individual decay rate. To really follow such dynamics, one resorts to a time-domain spectroscopy [11–13], in which a first "pulse" initiates the process and a second "pulse" probes it. Since the relevant time scale for such dynamics spans from attoseconds to femtoseconds and the relevant energy scale spans from 10^2 to 10^5 eV, X-ray attosecond bursts may be the choice to serve as the pump and the probe events. However, with the low photon flux of current soft X-ray attosecond sources ($\hbar\omega > 300$ eV) and the low absorption cross sections in this spectral range, it is currently impossible to both pump and probe these processes with attosecond X-ray pulses. To probe processes involving valence electrons, an ultrashort infrared pulse is often used to initiate the process, and a well-synchronized extreme ultraviolet (XUV) attosecond pulse probes it [14]. It is difficult to extend this scheme to excite inner-shell processes because of the large energy difference between inner-shell energies and the infrared photon energy. Excitation of inner-shell dynamics by laser-induced electron recollision might be the solution. Here, we show direct evidence for such excitations, as opposed to previous indirect evidence [15, 16]. Such an excitation process occurs in a sub-femtosecond time scale, and thus, provides the necessary "pump" step and might become the key for future "pump-probe" studies of inner-shell dynamics.

13.2 Experiments and Results

In our experiment, we focus a 12 fs, 1 mJ infrared radiation source ($\lambda_0 = 1.8$ μm) on a pulsed gas jet and observed the soft X-ray radiation from the interaction region. Because of the quadratic scaling of the ponderomotive energy, such infrared sources have already demonstrated the extension of HHG spectra toward the soft X-ray range ($\hbar\omega > 1$ keV) [15]. The infrared radiation source we have used here is based on an optical parametric amplifier, described in detail elsewhere [17]. The soft X-ray radiation that is coming from the excited atoms is a measure of the amount of excitation, but in order to separate it from the accompanying HHG radiation, we observed the soft X-ray radiation at a right angle to the infrared propagation direction. The infrared beam was focused on the gas target placed inside a vacuum chamber by a $f = 300$ mm CaF_2 lens to a spot size of about 65 μm FWHM, which results in a peak intensity of about 3×10^{15} W/cm^2. The corresponding ponderomotive energy is about 900 eV, enough to excite the K shell of neon and the L shell of krypton. For the gas target we used jets of neon and krypton from a pulsed nozzle (series 9 Parker nozzle, orifice diameter of 350 μm and backing pressures ranging from 1 to 10 bar), the pulsed nozzle was operated at 20 Hz to keep the pressure inside the vacuum chamber below 10^{-3} mbar. Soft X-ray spectra from the krypton and the neon atom were recorded by a silicon drift detector (Amptek XR-100SDD) with a 12.5 μm beryllium window, positioned 350 mm away from the interaction region. During the experiment, we firstly scanned the focus position along the gas propagation direction to find the best X-ray yield. We obtained the best X-ray yield when we placed the beam focus just next to the nozzle exit; all of the results were taken at this position.

Figure 13.1 shows the soft X-ray spectra coming from Ne and Kr targets. The spectrum shows the characteristic K-shell line from Ne and L-shell line from Kr, on top of a wider continuum. The continuum radiation from Kr extends up to ~2800 eV which is in good agreement with the abovementioned formula for the maximum possible kinetic energy of the recolliding electron: $K_{\max} = 3.17U_p = 2.96 \times 10^{-13}\, I_0\lambda^2$ (K is in eV, I_0 is in W/cm^2 and λ is in μm). We speculate that the origin of this continuum is either coming from the recombination radiation or from the bremsstrahlung radiation. The sharp cutoff at the lower energy end is due to absorption in the beryllium window.

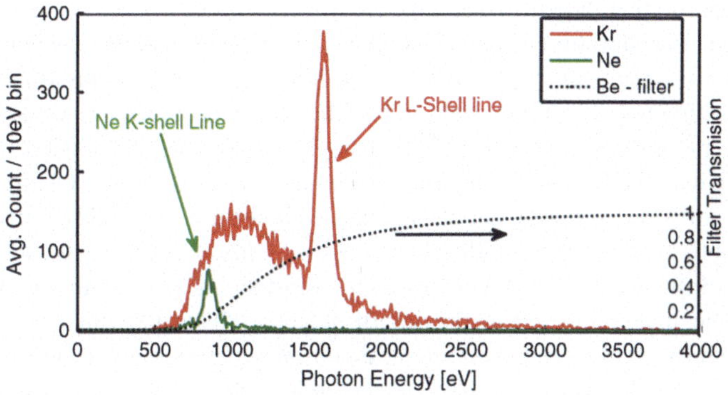

Fig. 13.1 The fluorescence spectra from the excited Ne (green line) and Kr (red line) atoms. The signal from the Ne is mainly from the K-shell transition with a weak continuum that stretches up to ≈1600 eV. In the Kr spectrum there is a sharp line, belonging to the L-shell transition, on top of a pronounced continuum. (The resolution of the SDD detector is not high enough to resolve the Lα, Lβ splitting.) The above Ne spectrum and the Kr spectrum were taken at different conditions. The Kr spectrum was taken during an integration time of 85 min and backing pressure of 4 bar; the Ne spectrum was taken during an integration time of 170 min and backing pressure of 10 bar. Also given in this figure is the transmission curve of the 12.5 μm thick beryllium window (black dotted line). Taking into account the repetition rate, integration time, and distance of the detector from the target, we estimate the total number of photons/pulse going into 4π as 10000 photons from Kr and 300 photons from Ne. The beryllium absorption is not included in the calculation

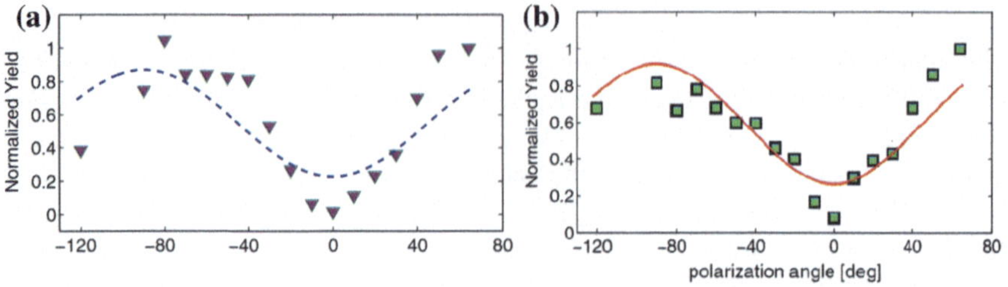

Fig. 13.2 The krypton L-shell fluorescence yield (**a**) and the continuum yield (**b**) as a function of the angle between polarization direction and detector direction. We fit the fluorescence yield to the $I(\theta)/I(\pi/2) = 1 - P\cos^2\theta$ formula (blue broken line, $P = 0.716$) and the continuum with the modified Sommerfeld formula which is given above (red solid line, $P = 0.78$). Measuring conditions: backing pressure of 4 bars, 5 min integration time for each point

The goal of this study is to show that the core-hole excitations we have observed are indeed coming from the recollision process and not from mere heating of the plasma. As a first step toward this goal, we followed the reasoning of [18] and tested the fluorescence and the continuum directionality. For that purpose, we used a $\lambda/2$ waveplate to rotate the polarization direction with respect to the position of the detector and measured both the X-ray fluorescence and the continuum yield as a function of that direction. We observed a minimum in both of them when the polarized field pointed toward the detector and a maximum when it was approximately perpendicular to that direction (see Fig. 13.2).

With the pulse duration of our IR source (only 2 cycle) and the gas densities we worked at 10^{17}–10^{18} cm^{-3}; we are not expecting the inverse bremsstrahlung (IB) and ATI heating to play an important role in the observed core-hole excitations. The dipole like radiation pattern (Fig. 13.2) strongly supports the recollision excitation mechanism over the IB and ATI heating processes. As a next step for testing whether we have recollision excitation or not, we checked how the X-ray yield depends on the drive's ellipticity, since the recollision process is highly sensitive to the polarization ellipticity of the drive [19]. According to the most simplified recollision model, in which the electron emerges from the deformed Coulomb barrier with a zero velocity, as the ellipticity gets larger and larger, the electron trajectories are pushed away from the parent ion and never come back to recollide with it. Therefore, it is common practice to check whether a process is coming from the recollision process or not by changing the drive polarization ellipticity. Figure 13.3a shows the x-ray yield from the Kr target as a function of the drive ellipticity.

Indeed, we can see a strong reduction in the X-ray yield as the ellipticity grows. However, as the ellipticity grows, the infrared peak intensity is reduced and leads to a reduced ionization rate which can partially explain the reduction of the photon yield. To check if the reduction in the photon yield is due to the lower ionization yield or due to the deflection of the returning electron, we compare again the results from the linear polarization drive and the circular polarization drive, this time we keep the peak electric field the same (see Fig. 13.3b). This test shows clearly that the main reduction

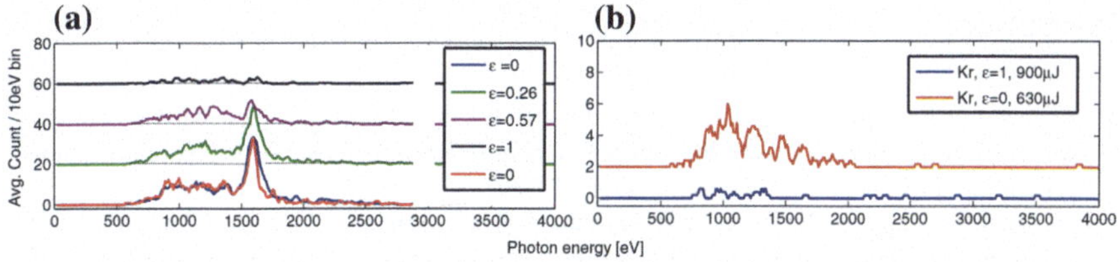

Fig. 13.3 **a** The X-ray radiation yield from the Kr atoms versus the drive ellipticity (backing pressure: 4 bar, integration time: 20 min per spectrum, laser energy 1050 μJ). **b** Comparison of the X-ray yield when using a linearly polarized or circularly polarized infrared drive while keeping the peak electric field the same (backing pressure 4 bar, integration time of 60 min for each spectrum)

in the X-ray yield is a result of the electron deflection by the circular polarization. Nevertheless, the signal does not completely disappear with circular polarization as one would expect from the most simplified recollision induced excitation model. This residual X-ray emission might have a connection to other recently reported findings from other groups. A nonsequential double ionization with circular polarization was reported by a few groups [20–23]. Mizuno et al. observed an extended tail in the spectrum of the photoelectrons from the interaction of a strong circular laser field with Kr atoms, such a tail was believed to be a signature for electron recollision. An explanation for all these findings might be given by an extended recollision model. In this extended model, the initial velocity of the electron just at the emergence time from the tunneled barrier is not necessarily equal to zero but has some distribution, both along the radiation polarization [24] as well as in the transverse direction [25, 26]. Some of these initial velocities allow for recollisions even with circularly polarized radiation [23, 27, 28]. Another option is a "shake-up" process, in which tunnel ionization results in simultaneous excitation of one or more of the remaining electrons [22]. It is also possible that the gas jet is thick enough to allow for fast ATI electrons to collide with atoms within the jet and excite them.

13.3 Conclusion

In conclusion, the dipole like radiation pattern (Fig. 13.2) and the strong dependency of the X-ray emission on the drive ellipticity (Fig. 13.3) are strong evidence supporting the recollision excitation mechanism over the IB and ATI heating processes [29]. Since the recollision excitation process occurs at the sub-femtosecond time scale, it opens the door for time-domain studies of electron dynamics in highly excited states where the recollision event initiates the excitation followed by a synchronized attosecond probe pulse.

References

1. T. Ditmire, E.T. Gumbrell, R.A. Smith, J.W.G. Tisch, D.D. Meyerhofer, M.H.R. Hutchinson, Spatial coherence measurement of soft x-ray radiation produced by high order harmonic generation. Phys. Rev. Lett. **77**, 4756–4759 (1996)
2. M. Bellini, C. Lyngå, A. Tozzi, M.B. Gaarde, T.W. Hänsch, A. L'Huillier, C.-G. Wahlström, Temporal coherence of ultrashort high-order harmonic pulses. Phys. Rev. Lett. **81**, 297–300 (1998)
3. P. Salières, A. L'Huillier, M. Lewenstein, Coherence control of high-order harmonics. Phys. Rev. Lett. **74**, 3776–3779 (1995)
4. Y. Mairesse, A. deBohan, L.J. Frasinski, H. Merdji, L.C. Dinu, P. Monchicourt, P. Breger, M. Kovačev, R. Taïeb, B. Carré, H.G. Muller, P. Agostini, P. Salières, Attosecond synchronization of high-harmonic soft x-rays. Science **302**, 1540–1543 (2003)
5. H. Niikura, H.J. Wörner, D.M. Villeneuve, P.B. Corkum, Probing the spatial structure of a molecular attosecond electron wavepacket using shaped recollision trajectories. Phys. Rev. Lett. **107**, 093004 (2011)
6. M. Kitzler, M. Lezius, Spatial control of recollision wavepackets with attosecond precision. Phys. Rev. Lett. **95**, 253001 (2005)
7. A. Baltuška, T. Udem, M. Uiberacker, M. Hentschel, E. Goulielmakis, C. Gohle, R. Holzwarth, V.S. Yakovlev, A. Scrinzi, T.W. Hänsch, F. Krausz, Attosecond control of electronic processes by intense light fields. Nature **421**, 611–615 (2003)
8. D. Shafir, Y. Mairesse, D.M. Villeneuve, P.B. Corkum, N. Dudovich, Atomic wavefunctions probed through strong-field light–matter interaction. Nat. Phys. **5**, 412–416 (2009)
9. P.B. Corkum, Plasma perspective on strong field multiphoton ionization. Phys. Rev. Lett. **71**, 1994–1997 (1993)
10. M. Drescher, M. Hentschel, R. Kienberger, M. Uiberacker, V. Yakovlev, A. Scrinzi, T. Westerwalbesloh, U. Kleineberg, U. Heinzmann, F. Krausz, Nature (London) **419**, 803 (2002)
11. F. Penent, J. Palaudoux, P. Lablanquie, L. Andric, R. Feifel, J.H.D. Eland, Phys. Rev. Lett. **95**, 083002 (2005)
12. T. Uphues, M. Schultze, M.F. Kling, M. Uiberacker, S. Hendel, U. Heinzmann, N.M. Kabachnik, M. Drescher, New J. Phys. **10**, 025009 (2008)
13. A.J. Verhoef, A.V. Mitrofanov, X.T. Nguyen, M. Krikunova, S. Fritzsche, N.M. Kabachnik, M. Drescher, A. Baltuška, New J. Phys. **13**, 113003 (2011)
14. F. Krausz, M. Ivanov, Rev. Mod. Phys. **81**, 163 (2009)
15. G. Marcus, W. Helml, X. Gu, Y. Deng, R. Hartmann, T. Kobayashi, L. Strueder, R. Kienberger, F. Krausz, Phys. Rev. Lett. **108**, 023201 (2012)
16. A.D. Shiner, B.E. Schmidt, C. Trallero-Herrero, H.J. Worner, S. Patchkovskii, P.B. Corkum, J.-C. Kieffer, F. Legare, D.M. Villeneuve, Nat. Phys. **7**, 464 (2011)
17. C. Zhang, P. Wei, Y. Huang, Y. Leng, Y. Zheng, Z. Zeng, R. Li, Z. Xu, Opt. Lett. **34**, 2730 (2009)
18. S. Dobosz, M. Lezius, M. Schmidt, P. Meynadier, M. Perdrix, D. Normand, J.-P. Rozet, D. Vernhet, Phys. Rev. A **56**, R2526 (1997)
19. K.S. Budil, P. Salières, M.D. Perry, A. L'Huillier, Phys. Rev. A **48**, R3437 (1993)
20. G.D. Gillen, M.A. Walker, L.D. Van Woerkom, Phys. Rev. A **64**, 043413 (2001)
21. C. Guo, G.N. Gibson, Phys. Rev. A **63**, 040701 (2001)
22. W.A. Bryan et al., Nat. Phys. **2**, 379 (2006)
23. F. Mauger, C. Chandre, T. Uzer, Phys. Rev. Lett. **105**, 083002 (2010)
24. A.N. Pfeiffer, C. Cirelli, A.S. Landsman, M. Smolarski, D. Dimitrovski, L.B. Madsen, U. Keller, Phys. Rev. Lett. **109**, 083002 (2012)
25. M.Y. Ivanov, M. Spanner, O. Smirnova, J. Mod. Opt. **52**, 165 (2005)
26. L. Arissian, C. Smeenk, F. Turner, C. Trallero, A.V. Sokolov, D.M. Villeneuve, A. Staudte, P.B. Corkum, Phys. Rev. Lett. **105**, 133002 (2010)
27. N.I. Shvetsov-Shilovski, S.P. Goreslavski, S.V. Popruzhenko, W. Becker, Phys. Rev. A **77**, 063405 (2008)

28. X. Wang, J.H. Eberly, New J. Phys. **12**, 093047 (2010)
29. Yunpei Deng, Zhinan Zeng, Zhengmao Jia, Pavel Komm, Yinhui Zheng, Xiaochun Ge, Ruxin Li, Gilad Marcus, Ultrafast excitation of an inner-shell electron by laser-induced electron recollision. Phys. Rev. Lett. **116**, 073901 (2016)

Chapter 14
User-Oriented High-Harmonic Source at ELI Beamlines

O. Finke, O. Hort, M. Albrecht, V. E. Nefedova, D. D. Mai, F. Giambruno, S. Reyne, L. Poletto, F. Frassetto, J. Gautier, S. Sebban, R. Antipenkov, F. Batysta, R. Boge, J. Naylon, Z. Hubka, T. Green, P. Bakule and J. Nejdl

Abstract We present experimental setup for user-oriented high-harmonic source with a description of different focusing and interaction geometries. Infrared rejection system and resulting extreme ultraviolet beam characterization are also described. Beamline is designed to be driven with 100 mJ, <20 fs pulses centered at 830 nm at 1 kHz repetition rate. Results from beamline commissioning with commercial 1 kHz, 5 mJ and 40 fs laser system and testing with in-house developed laser L1 are included.

14.1 Introduction

High harmonic generation provides main source of highly coherent, ultrashort extreme ultraviolet (XUV) and soft X-ray radiation. High coherence of the generated beam is an important parameter for coherent diffractive imaging [1]. Short wavelengths and large bandwidth of the XUV radiation enable generating pulses in lengths of tens of attosecond (1 as $= 10^{-18}$ s) which can be used in probing ultrafast molecular and atomic dynamics [2, 3].

Such source of coherent XUV radiation, based on HHG in gases, oriented at user research, is actively being developed and implemented at ELI Beamlines facility with

O. Finke (✉) · O. Hort · M. Albrecht · V. E. Nefedova · D. D. Mai · J. Gautier · S. Sebban · R. Antipenkov · F. Batysta · R. Boge · J. Naylon · Z. Hubka · T. Green · P. Bakule · J. Nejdl
ELI Beamlines, Institute of Physics ASCR, Prague, Czech Republic
e-mail: Ondrej.Finke@eli-beams.eu

O. Finke · M. Albrecht · V. E. Nefedova · F. Batysta · Z. Hubka
FNSPE, Czech Technical University in Prague, Prague, Czech Republic

F. Giambruno · S. Reyne
ARDOP Industrie, Cité de la Photonique, Pessac, France

L. Poletto · F. Frassetto
Institute of Photonics and Nanotechnologies, University of Padua, Padua, Italy

J. Gautier · S. Sebban
Laboratoire d'Optique Appliquée, Palaiseau, France

M. Kozlová and J. Nejdl (eds.), *X-Ray Lasers 2018*, Springer Proceedings in Physics 241, https://doi.org/10.1007/978-3-030-35453-4_14

accompanied research activities [4–6]. Here we present description of the system's geometry, its diagnostics and results obtained within the commissioning of the system with a commercial laser driver.

14.2 High-Harmonic Beamline

The high-harmonic beamline is designed as a versatile source offering wide variety of generating options to fulfill users' requirements. The system consists of four large vacuum chambers (VCs), XUV Monochromator, and one small vacuum chamber, called filter box (FB) (Fig. 14.1).

14.2.1 Beamline Description

Chambers are connected with vacuum tubes DN250 to adopt large driving beam. The chamber VC1 is movable allowing change of focusing geometry. Options for focusing are 2.5, 5, 8, 12.5, 20, and 25 m resulting in f-numbers 90, 180, 290, 450, 730, and 900, respectively.

The laser beam enters the system into chamber VC2 and is folded and focused through chambers VC3 and VC1 to a gas cell located in chamber VC2. Exact route

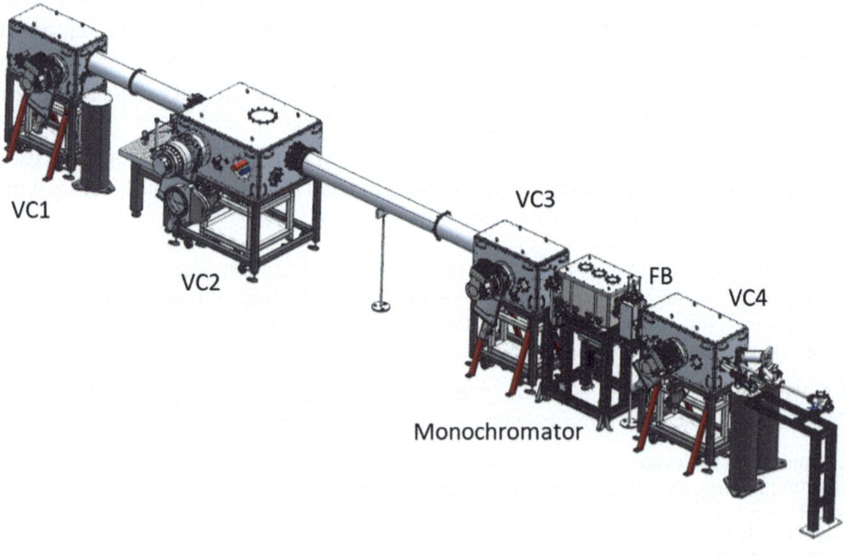

Fig. 14.1 Scheme of the harmonic beamline in short focusing configuration. From left VC1–VC2–VC3–Monochromator–FB–VC4

Table 14.1 Diffraction gratings options considered for the monochromator and their corresponding spectral ranges

Gratings (lines/mm)	86	158	600	985
Spectral region (eV)	10–28	25–54	51–98	86–121

of the IR beam depends on required focusing geometry. Focal spot of the beam can be moved relative to the gas cell using 0.5 m translation stage in VC1. After passing through the gas cell, driver beam continues to chamber VC3 together with the generated harmonic beam. VC3 houses IR rejection system consisting of three infrared antireflective-coated mirrors under incidence angle of 87°, 84°, and 87°. IR light is transmitted through the mirrors, in contrast to the XUV beam which is reflected while keeping its original beam axis. Removable mirror is placed after the rejection system to direct harmonic beam into VUV ellipsometry setup [7] or let the beam propagate to XUV monochromator designed to reduce XUV bandwidth using grazing incidence collimating toroidal mirror, diffraction grating (Table 14.1), and another toroidal mirror focusing the angularly dispersed beam on a slit or let the beam continue without changes. Monochromator is followed by two sets of thin metallic filters which reject the rest of the IR beam and transmit only the harmonic beam.

Generated beam then propagates into VC4 where it is diagnosed or let freely continue into user Multipurpose Atomic molecular and optical science and Diffraction imaging Chamber (MAC) based on the design of the CAMP station at FLASH [8].

14.2.2 XUV Diagnostics

XUV diagnostics consists of XUV photodiode, XUV spectrometer, and wavefront sensor. Position of movable translation stage at the beginning of the chamber VC4 defines the diagnostics used. Translation stage houses photodiode, flat mirror to reflect the beam to a wavefront sensor, and a toroidal mirror for the XUV spectrometer.

Energy of XUV beam is measured using charge generated by calibrated XUV photodiode. Novel system of online energy measurement is also being developed based on online measurement of charge generated by the harmonic beam passing through the thin metallic filters. Preliminary results were measured using 423 nm thin aluminium filter. Bias voltage of −80 V was applied onto the filter and signal-to-noise ratio was furthermore improved using amplifier with factor of 10^6. Different energies of harmonic pulses were achieved by varying gas pressure. Charge measured from the filter was compared to the one measured by the calibrated diode.

Harmonic beam spectrum is characterized using XUV spectrometer consisting of a toroidal mirror, a motorized slit, a reflection grating, and a XUV CCD. Generating

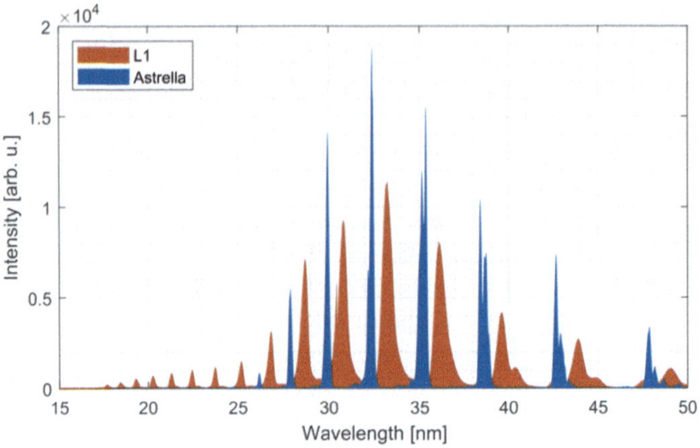

Fig. 14.2 Spectrum of argon generated in short focusing configuration ($f = 2.5$ m) using L1 laser system (red) and Astrella laser system (blue)

medium is horizontally imaged onto the slit. Concave grating with variable line spacing then images the slit in the horizontal direction onto the CCD chip. Astigmatism caused by the concave spherical grating is compensated by correctly designed radius of the toroidal mirror. Usage of toroidal mirror furthermore improves signal-to-noise ratio of the whole system by increasing photon flux. Two gratings are available with groove density of 1200 lines/mm and 600 lines/mm which corresponds to spectral range of 5–60 nm and 20–120 nm, respectively. Figure 14.2 shows spectra measured in argon during the tests performed with the L1 laser driver (1 mJ of pulse energy) compared to spectrum from commercial Astrella laser system.

The wavefront of the XUV beam is measured using a Hartmann-type wavefront sensor. The grid of the wavefront sensor consists of 60×60 holes with 210 μm pitch and 43.5 mm distance from the CCD chip. The camera itself has pixel size of 13.5 μm and 1024×1024 pixels. Final sensitivity of the system is better than $\lambda/5$ down to 10 nm wavelength. The camera is mounted on the back of VC4.

14.2.3 Future Upgrades on the HHG Beamline

Development of multiple upgrades is currently in various stages. Most advanced is implementation of XUV monochromator which is being currently tested. Beamline will be also capable to operate with two-color driver and will be able to provide two parallel harmonic beams from two separate gas cells. For advanced pump-probe experiment it is planned to integrate synchronized THz coherent beam, Kα radiation, and OPA tuneable laser.

Tab. 14.2 Table of measured XUV beam properties with commercial laser system (5 mJ, 40 fs) and calculated properties with final L1—Allegra laser parameters

Laser system	Gas	XUV wavelength (nm)	Driver F#	XUV pulse energy (μJ)	XUV divergence (mrad)
5 mJ, 40 fs	Xenon	≥51	280	0.05	0.6
100 mJ, 20 fs			1430	2	0.1
5 mJ, 40 fs	Argon	≥32	120	0.005	0.8
100 mJ, 20 fs			625	0.2	0.15
5 mJ, 40 fs	Neon	≥13.5	87	5×10^{-4}	0.48
100 mJ, 20 fs			444	0.02	0.09
5 mJ, 40 fs	Helium	≥10	75	5×10^{-4}	0.4
100 mJ, 20 fs			380	0.02	0.07

14.3 Measured and Expected Beam Parameters

Presented beamline was designed for the in-house developed laser L1-Allegra of the ELI Beamlines facility [9]. Laser is planned to operate at 1 kHz providing sub-20 fs pulses centered at 830 nm with energy of 100 mJ. Functionality of the beamline was tested and confirmed with commercial laser system providing 40 fs pulses with 5 mJ energy and central wavelength of 810 nm. This system also operates at 1 kHz frequency. First tests of the beamline with L1-Allegra system at 1 mJ pulse energy were also performed (Fig. 14.2). Due to shorter driver pulses, bandwidth of the individual peaks is wider and XUV cutoff moves to shorter wavelengths compared to spectrum generated with commercial laser system.

In Table 14.2, we present measured XUV beam parameters from commissioning of the beamline where Rayleigh range, comparable to optimal medium length, was 6.4 cm, 1.22 cm, and 0.55 cm for HHG in xenon, argon, and neon, respectively. Expected XUV beam parameters with full power L1 driver (100 mJ per pulse) are also included.

14.4 Conclusion

We presented versatile user-oriented high-harmonic beamline built at ELI Beamlines facility. This source currently runs on 1 kHz Ti:Sapphire CPA laser system with 5 mJ, 40 fs pulses centered at 810 nm and in near future will benefit from the OPCPA laser with 100 mJ, 20 fs centered at 830 nm.

Whole beamline was presented with focus on beam path, different geometries and XUV beam diagnostics. Furthermore, system is now available for users which require robust coherent XUV source. Users can utilize multipurpose MAC station and a magneto-optical VUV ellipsometer which are under implementation.

More details of the beamline setup and description of future upgrades are described in [10].

Acknowledgements Project LG1606 was financially supported by the Ministry of Education, Youth and Sports as part of targeted support from the National Programme of Sustainability II and supported by the project Advanced research using high-intensity laser-produced photons and particles (CZ.02.1.01/0.0/0.0/16_019/0000789) from European Regional Development Fund (ADONIS). We also acknowledge project no. CZ.1.07/2.3.00/20.0279 that was co-financed by the European Social Fund and the state budget of the Czech Republic.

References

1. Z. Bosheng, Coherent diffractive imaging microscope with tabletop high harmonic EUV source. *Proceedings SPIE Metrology, Inspection, and Process Control for Microlithography XXVII* (2013), p. 86810H
2. K. Klünder, Probing single-photon ionization on the attosecond time scale. Phys. Rev. Lett. **106**, 143002 (2011)
3. R. Pazourek, Attosecond chronoscopy of photoemission. Rev. Mod. Phys. **87**, 765 (2015)
4. V.E. Nefedova, Efficiency control of high-order harmonic generation in gases using driving pulse spectral features. Appl. Phys. Lett. **113**, 191101 (2018)
5. V.E. Nefedova, Determination of the spectral variation origin in high-order harmonic generation in noble gases. Phys. Rev. A **98**, 033414 (2018)
6. V.E. Nefedova, Development of a high-flux XUV source based on high-order harmonic generation. J. Electron Spectrosc. Relat. Phenom. **220**, 9 (2017)
7. S. Espinoza, User oriented end-station on VUV pump-probe magneto-optical ellipsometry at ELI beamlines. Appl. Surf. Sci. **421**, 378–382 (2017)
8. L. Strüder, Large-format, high-speed, x-ray pnccds combined with electron and ion imaging spectrometers in a multipurpose chamber for experiments at 4th generation light sources. Nucl. Instrum. Methods Phys. Res. Sect. A. **614**, 483–496 (2010)
9. F. Batysta, Broadband OPCPA system with 11 mJ output at 1 kHz, compressible to 12 fs. Opt. Express **24**, 17843 (2016)
10. O. Hort, High-flux source of coherent XUV pulses for user application. Opt. Express **27**, 8871 (2019)

Chapter 15
Enhancement of Harmonics Generated in Modulated Indium Laser-Plasmas: Experiment and Theory

S. Y. Stremoukhov, R. A. Ganeev and A. V. Andreev

Abstract We compare experimental and theoretical studies of the quasi-phase matching (QPM) of high-order harmonic generation (HHG) during propagation of two-color pulses through the indium multi-jet plasmas. Role of the sizes of plasma jets and distance between them on the HHG efficiency at QPM conditions is analyzed. The results of the QPM conditions' studies provide the reliable interpretation of this phenomenon.

15.1 Introduction

Different regimes of laser-plasma interaction attract interest due to possibility in conversion of longer wavelength laser sources toward extreme ultraviolet (XUV) region. Various methods are used to amend this process [1–3]. The latter approach, though attractive in the case of ablation of extended targets, has a significant drawback related with the phase mismatch between interacting waves of driving pulse and harmonic when the length of medium exceeds so-called coherence length for the generated harmonic order.

In this paper, we analyze different patterns of experimentally observed modifications of harmonic spectra at the conditions of QPM in indium multi-jet plasmas. In particular, we analyze the role of the sizes of plasma jets and distance between them

S. Y. Stremoukhov (✉)
National Research Centre "Kurchatov Institute", Pl. Akademika Kurchatova 1, Moscow 123182, Russia
e-mail: sustrem@gmail.com

S. Y. Stremoukhov · A. V. Andreev
Faculty of Physics, Lomonosov Moscow State University, Leninskie Gory 1/2, Moscow 119991, Russia

R. A. Ganeev
Faculty of Physics, Voronezh State University, University Square 1, Voronezh 394006, Russia

The Guo China-US Photonics Laboratory, State Key Laboratory of Applied Optics, Changchun Institute of Optics, Fine Mechanics and Physics, Chinese Academy of Sciences, Changchun 130033, China

M. Kozlová and J. Nejdl (eds.), *X-Ray Lasers 2018*, Springer Proceedings in Physics 241, https://doi.org/10.1007/978-3-030-35453-4_15

on the HHG efficiency at QPM conditions. We demonstrate the two-color induced QPM at different regimes of laser-plasma interaction. We also compare the experimental data with our calculations based on the model described in [4]. More detailed investigation of the QPM of the HHG during propagation of two-color pulses through the multi-jet plasmas are presented in [5].

15.2 Results

Figure 15.1 shows results of the group harmonics enhancement in the indium plasma. Experimental results have been obtained with the method, presented in [5, 6]. In these HHG studies, the 810 nm, 50 fs, 10 Hz and 810 nm, 340 ps, 10 Hz pulses from Ti:Sapphire laser were used as the driving and heating radiation. We also used the 70 fs, near-infrared (NIR) pulses from optical parametric amplifier for pumping the modulated plasmas and harmonic generation. The 340 ps pulses were focused using 300-mm focal length cylindrical lens on the surface of the ablating target placed in the vacuum chamber to form the homogeneous extended plasma (l = 5 mm). The intensity of these pulses on the target surface was up to $I = 4\times10^9$ W cm^{-2}. The plasma area was 5×0.08 mm^2. To produce multi-jet plasmas the multi-slit mask (MSM) was installed between cylindrical lens and target. By tilting this mask at different angles we were able to produce different numbers of equidistant plasma jets. The current results have been obtained for two-color laser field formed by the radiation at 1310 and 655 nm (or 1300 and 650 nm, or 1320 and 660 nm) wavelength. Figure 15.1a (from top to bottom) shows the part of the generated spectra obtained for homogeneous plasma having 11-, 9-, and 8-jets. It is clearly seen that the spectra obtained from the perforated plasma (with jets) contain significantly enhanced groups of harmonics (centered at 35, 32, and 27 orders for 11-, 9-, and 8-jets, with the sizes of singe plasma jets of 0.23, 0.28, and 0.31 mm, respectively). Figure 15.1a also demonstrates strongly enhanced emission near 21st harmonic which corresponds to transitions with high oscillation strength (gf = 11) [5], but the explanation of this effect is not the target of the current study.

To theoretically support these results, we have simulated cases described above by using the interference model described in [4]. Within this model, the nonlinear medium is simulated by 1D chain of atoms oriented along the propagation direction of laser pulses. The response of each atom in the chain is calculated using the non-perturbative theory [7, 8]. The two-color laser field parameters that were chosen for the single-atom response calculations on one hand coincide with the ones used in the experiments, and on the other hand depend on the position of the atoms in the chain. Generated harmonic radiation from each atom is calculated in the far-field zone by a numerical summation over chain of atoms [4, 9]. As a result, the spatial distribution of generated radiation is calculated, which is integrated for analysis of overall intensity of the harmonic radiation. Figure 15.1b shows the results of these studies and demonstrates the good agreement between the experimental results and theoretical calculations.

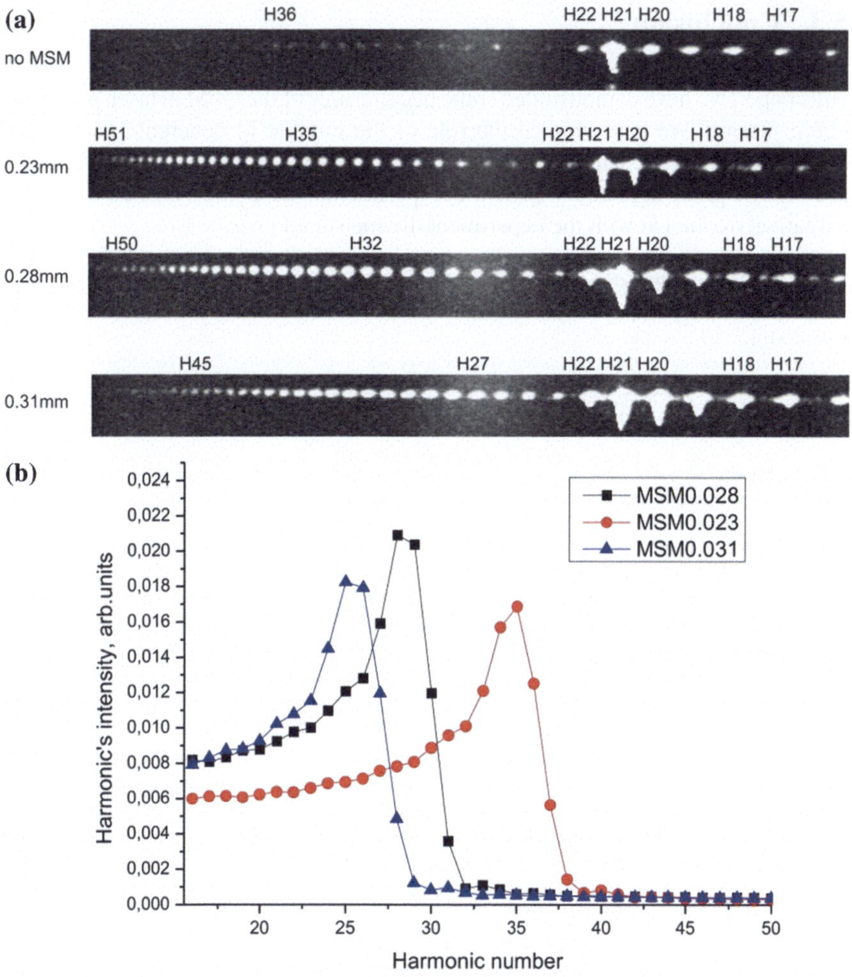

Fig. 15.1 **a** Raw images of harmonic spectra from the multi-jet indium laser plasma produced without (from top to bottom) and using tilted MSM with single plasma jet sizes of 0.23, 0.28, and 0.31 mm pumped by two-color (1310 + 655 nm of imperforated plasma, 1300 + 650 nm for perforated with 0.23 mm jets and 1320 + 660 nm for perforated with 0.28 and 0.31 mm jets), 70 fs pulses. Maximally enhanced harmonics (H35, H32, and H27) are determined for each of these perforated plasma configurations. Similar conditions of plasma ablation using the same energy fluency of heating pulses and data collection times are applied for these four cases. **b** Harmonic distribution from multi-jet indium plasma in the case of propagation of two-color driving pulses (with exactly the same parameters as those from corresponding experiments) through the plasma of ~5 mm length containing the jets with the sizes equal to 0.28 mm (black squares), 0.23 mm (red circles), 0.31 mm (blue triangles)

15.3 Conclusion

In this paper, we have demonstrated some peculiarities of the QPM in laser-produced indium plasma. We have studied the role of the number of coherent zones, sizes of plasma jets and distance between them. The results of theoretical simulations have shown good agreement with the experimental data: the calculated spectra qualitatively coincide with the experimentally measured ones.

Acknowledgements R.A.G. acknowledges the support from H. Kuroda. The theoretical part of the work was partially supported by the Russian Foundation for Basic Research under Projects Nos. 18-02-40014.

References

1. R.A. Ganeev, Generation of high-order harmonics of high-power lasers in plasmas produced under irradiation of solid target surfaces by a prepulse. Phys. Usp. **52**, 55 (2009)
2. Z. Abdelrahman, M.A. Khokhlova, D.J. Walke, T. Witting, A. Zair, V.V. Strelkov, J.P. Marangos, J.W.G. Tisch, Chirp-control of resonant high-order harmonic generation in indium ablation plumes driven by intense few-cycle laser pulses. Opt. Express **26**, 15745 (2018)
3. R.A. Ganeev, V.V. Strelkov, C. Hutchison, A. Zaïr, D. Kilbane, M.A. Khokhlova, J.P. Marangos, Experimental and theoretical studies of two-color-pump resonance-induced enhancement of odd and even harmonics from a tin plasma. Phys. Rev. A **85**, 023832 (2012)
4. S.Y. Stremoukhov, A.V. Andreev, Quantum-mechanical elaboration for the description of low- and high-order harmonics generated by extended gas media: prospects to the efficiency enhancement in spatially modulated media. Laser Phys. **28**, 035403 (2018)
5. R.A. Ganeev, S.Y. Stremoukhov, A.V. Andreev, Quasi-phase matching of harmonics generating in laser plasmas: experiment and theory. J. Opt. Soc. Am. B
6. R.A. Ganeev, M. Baba, M. Suzuki, H. Kuroda, High-order harmonic generation from silver plasma. Phys. Lett. A **339**, 103 (2005)
7. A.V. Andreev, S.Y. Stremoukhov, O.A. Shoutova, Light-induced anisotropy of atomic response: prospects for emission spectrum control. Eur. Phys. J. D **66**, 16 (2012)
8. S. Stremoukhov, A. Andreev, B. Vodungbo, P. Salières, B. Mahieu, G. Lambert, Origin of ellipticity of high-order harmonics generated by a two-color laser field in the cross-polarized configuration. Phys. Rev. A **94**, 013855 (2016)
9. S. Stremoukhov, A. Andreev, Quantum-mechanical fingerprints in generation of elliptical terahertz radiation by extended media interacting with two-color laser field. J. Opt. Soc. Am. B **34**, 232 (2017)

Chapter 16
Imaging Plate Absolute Calibration in the XUV Wavelength Range

P. Gajdoš, M. Kozlová and M. Krůs

Abstract The imaging plate (IP) represents a versatile diagnostics, which is resistant to electromagnetic pulses emerging during the high-power laser pulse interaction with a solid target. The IPs as a radiation recorder have advantages including reusable, high dynamic range, large active area. Nonetheless, the absolute calibration data of the IPs efficiency in the XUV wavelength range are missing up to now. At PALS Research Center, we performed the absolute calibration of the BAS-TR IPs being sensitive in the XUV wavelength range. The IPs were calibrated by cross-calibration with calibrated CCD camera. The XUV pulses were generated via high-order harmonic mechanism; wavelengths of these pulses were in the range of 20–40 nm. Moreover, the absolute efficiency curve as a function of the XUV radiation wavelength was determined.

16.1 Introduction

Imaging plates are two-dimensional, position sensitive detectors which can be used for detection of various types of radiation. By the reason of their immunity to electromagnetic pulses, they could be utilized in various high-power laser-plasma experiments [1, 2]. Fuji Biological Analysis System provides three types of IPs that are mostly used for detection. They consist of 3 or 4 layers: magnetic, support, phosphor, and protective. The BAS-TR IP, which does not have protective layer, can be used for XUV radiation detection because XUV light would not be able to penetrate through such a layer. IPs detection system is based on photostimulated phosphor ($BaF(Br,I):Eu^{2+}$). When IP is exposed to XUV radiation, a photoelectron is ejected into the conduction band and trapped in a lattice defects called F- or color-centers, which originate from a manufacturing process [3]. After that, the irradiated IP has to

P. Gajdoš (✉) · M. Kozlová · M. Krůs
Institue of Plasma Physics CAS, Za Slovankou 3, Prague 8 182 00, Czech Republic
e-mail: gajdos@ipp.cas.cz

M. Kozlová · M. Krůs
Institute of Physics CAS, Na Slovance 2, Prague 8 182 21, Czech Republic

M. Kozlová and J. Nejdl (eds.), *X-Ray Lasers 2018*, Springer Proceedings in Physics 241, https://doi.org/10.1007/978-3-030-35453-4_16

be scanned in the IP reader, where the reading laser light triggers the photoelecton recombination and releases photons in photostimulated luminescence (PSL) process [4, 5].

16.2 Experimental Setup

The calibration was performed by high-order harmonics generated by femtosecond Ti:sapphire laser available at PALS laboratory [6]. Within this measurement, Ti:sapphire laser delivered 300 mJ pulses with 50 fs duration at central wavelength of 808 nm. For the calibration in the XUV range of the electromagnetic spectrum, i.e., 20–40 nm, the 23rd–35th high-order harmonics were used. These harmonics were recorded by the flat-field spectrometer, which consisted of a curved grating with 1200 lines per mm and a detector (CCD or IP) (Fig. 16.1). An aluminum foil was used to filter HHG beam from laser beam.

Typical HHG spectrum accumulated over 250 shots recorded with CCD is shown in Fig. 16.2 and the analogous spectrum accumulated over 10,000 shots obtained from IP is in Fig. 16.3. The signal accumulation was performed to minimize measurement uncertainties, i.e., to maximally utilize the detector dynamic range, and to minimize shot-to-shot variations. Experimental conditions of high-order harmonics generation were kept identical for both detectors (CCD and IP), therefore irradiation conditions were also unchanged.

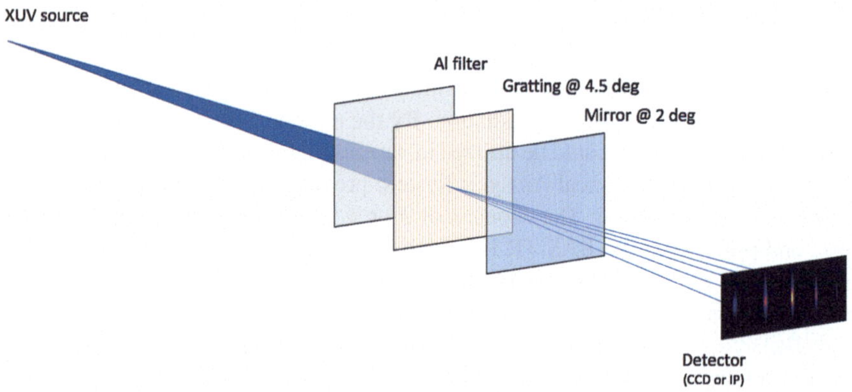

Fig. 16.1 Schematic sketch of the experimental setup. The XUV source is produced via high-order harmonic mechanism. HHG beam is filtered from laser beam using an aluminum foil. Subsequently, HHG beam is diffracted by a curved grating and detected by CCD or IP

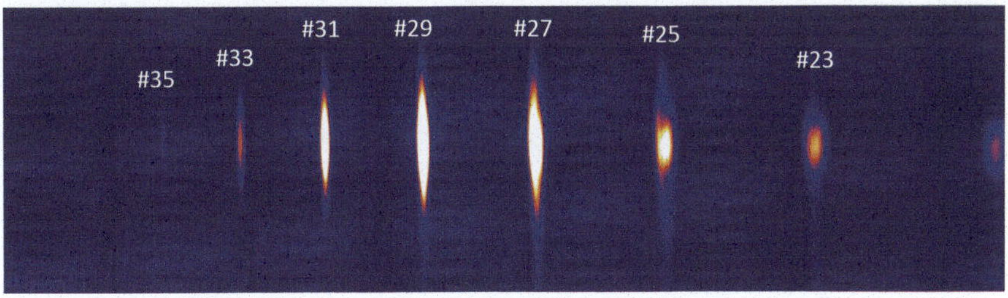

Fig. 16.2 Typical HHG spectrum recorded over 250 shots on CCD

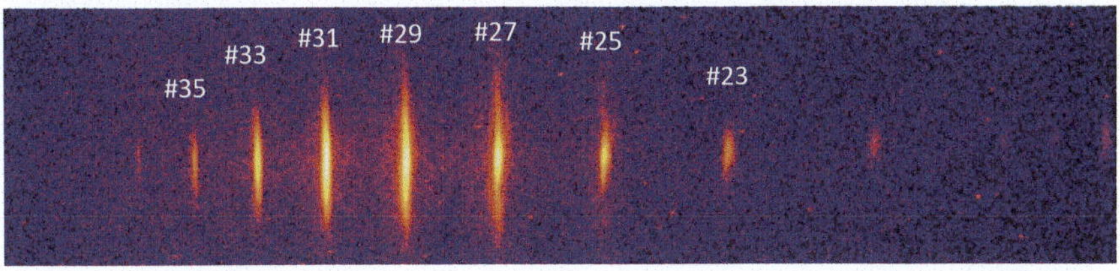

Fig. 16.3 Typical HHG spectrum accumulated over 10,000 shots retrieved from IP

16.3 Imaging Plate Absolute Calibration

The absolute calibration was provided by cross-calibration with CCD camera which allows to determine the absolute number of photons. Parameters of the camera were taken into account for calculation of this number, i.e., quantum efficiency of CCD chip, which varies from approximately 83 to 70 % in wavelength range of 20–40 nm, camera settings (high sensitivity, gain set to 2) and energy of 3.65 eV needed for creation of electron–hole pair in silicon CCD chip.

The imaging plate was scanned with BAS IP 1800-II reader with the following parameters: resolution $R = 50$ μm, sensitivity $S = 4000$, latitude $L = 5$, and gradation $G = 65{,}536$. The signal obtained from IP in QL units (image bits) was converted to PSL units using the following formula, which was given by the scanner producer.

$$\mathrm{PSL} = \left(\frac{R}{100}\right)^2 \left(\frac{4000}{S}\right) 10^{L\left(\frac{QL}{G}-\frac{1}{2}\right)}$$

The absolute spectral calibratIon of IP in the XUV range in terms of PSL per number of photons is shown in Fig. 16.4.

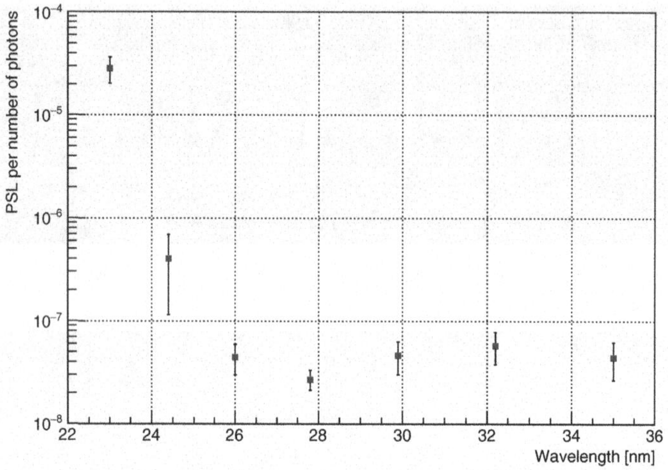

Fig. 16.4 Spectral calibration of IP in terms of PSL per number of photons

16.4 Conclusions

The absolute calibration of imaging plate FUJI BAS-TR was performed for XUV radiation in wavelength range of 20–40 nm. For the first time, it was accomplished in this wavelength region. The XUV pulses were generated via high-order harmonics produced by femtosecond Ti:sapphire laser at PALS laboratory. The flat-field spectrometer was used to record high-order harmonics. The IPs were calibrated by cross-calibration with CCD camera. The absolute efficiency curve in terms of PSL units per number of photons as a function of XUV wavelength was determined.

Acknowledgements The results of the Project LM2015083 were obtained with the financial support of the Ministry of Education, Youth and Sports within targeted support of Large infrastructures. Supported by Grant agency of the Czech Republic project number 18-27340S.

References

1. B.R. Maddox et al., Rev. Sci. Instrum. **82**, 023111 (2011)
2. N. Izumi et al., Rev. Sci. Instrum. **77**, 10E325 (2006)
3. Y. Amemiya, Synchrotron Radiat. **3**21, News (1990)
4. A. Curcio et al., J. Instrum. **11**, C05001 (2016)
5. A.L. Meadowcroft et al., Rev. Sci. Instrum. **79**, 113102 (2008)
6. V.E. Nefedova et al., App. Phys. Lett. **113**, 191101 (2018)

Part III
X-Rays from Relativistic Electron Beams

Chapter 17
Betatron X/γ-Ray Radiation from Wakefield-Accelerated Electrons Wiggling in Laser Fields

Y. F. Li, J. Feng, D. Z. Li, J. H. Tan, K. Huang, J. G. Wang, M. Z. Tao and L. M. Chen

Abstract Betatron radiation based on laser-plasma accelerator has properties of collimated, ultrafast, small sources size, high brightness, and broad energy spectrum. These features make it suitable for many applications, such as phase-contrast imaging or X-ray absorption spectroscopy. In this paper, we present two methods to improve the photon energy and photon number of betatron radiation. As a result, high-brightness photon sources with high yield, adjustable photon energy range from hard X-ray to gamma ray and polarization were generated. The main reason for the enhancement is due to the accelerated electrons efficiently wiggling in a laser field.

17.1 Introduction

Laser-plasma accelerators (LPAs) were proposed by T. Tajima and J. M. Dawson [1] in 1979. With the development of high-power laser technologies [2] and deep understanding of acceleration physical processes [3], now, LPAs can produce electron beams with high energy [4], high quality [5, 6], and large current intensity [7, 8]. These high-energy electron beams can be used to generate collimated, ultrafast, high-brightness x/gamma ray sources [9].

One important X-ray source based on LPAs is betatron-like radiation [10], which has potential for many applications [11]. In blowout/bubble regime [3], electrons injected into the cavity off-axis would be accelerated in the longitudinal direction and also transversally wiggled due to transverse restoring force produced by the

Y. F. Li · J. Feng · J. H. Tan · K. Huang · J. G. Wang · M. Z. Tao · L. M. Chen (✉)
Beijing National Laboratory of Condensed Matter Physics, Institute of Physics, CAS, Beijing 100190, China
e-mail: lmchen@iphy.ac.cn

D. Z. Li
Institute of High Energy Physics, CAS, Beijing 100049, China

L. M. Chen
School of Physical Sciences, University of Chinese Academy of Sciences, Beijing 100049, China

IFSA Collaborative Innovation Center, Department of Physics and Astronomy, Shanghai Jiao Tong University, Shanghai 200240, China

M. Kozlová and J. Nejdl (eds.), *X-Ray Lasers 2018*, Springer Proceedings in Physics 241, https://doi.org/10.1007/978-3-030-35453-4_17

transverse electric field, resulting in collimated beams of X-ray radiation, i.e., betatron radiation. The physics of betatron radiation in this scheme is well understood [9, 12], and can be directly compared to a synchrotron emission in the wiggler regime. The radiation features depend on strength parameter $K = \gamma\theta_{\max} = \gamma\omega_\beta r_\beta/c$, where γ is electron energy, $\omega_\beta = \omega_p/(2\gamma)^{1/2}$ is betatron frequency, ω_p is plasma frequency, and r_β is oscillation amplitude. In the wiggler regime, corresponding to $K \gg 1$, a synchrotron spectrum with closely spaced harmonics and critical energy $E_c = 3\hbar K\gamma^2\omega_\beta$ is generated and the radiation is confined to a cone with opening angle $\theta \approx K/\gamma$. The photon number emitted per electron per period for wiggler regime is $N_O = (5\pi/3^{1/2})\alpha K$, where $\alpha = e^2/(\hbar c) = 1/137$ is the fine structure constant. The total radiated photon number from an electron bunch with N_e electrons and N_β oscillation periods could be incoherently summed, i.e., $N_{total} = N_e N_\beta N_O$.

The radiated photon energy in this scheme can easily reach the range of X-ray or even gamma ray due to microns scale betatron period. In addition, betatron X-ray radiation is naturally synchronized with the driver laser (fs synchronization accuracy), which is very suitable for ultrafast pump-probe experiments. Since firstly observation of laser-driven betatron radiation in 2004 [10], many experiments have been carried out to measure and characterize this radiation sources worldwide, and demonstrated its small source size (a few microns) [13], small divergence angle (tens of milliradian) [14], ultrashort pulse duration [15], energy range from soft X-ray to γ ray [16, 17], high brightness [18, 19], and polarized X-ray [20, 21]. Relying on these merits, many applications with betatron radiation sources have been evaluated and carried out, such as X-ray phase-contrast imaging [22], computer tomography [23, 24], X-ray absorption spectroscopy [25, 26], or radiography of laser-driven shock waves [27].

Photon flux and operation energy range are two critical factors of light sources for applications. Therefore, improving accelerated electrons' energy, charge, oscillation amplitude, and frequency is very important for betatron radiation. In present, most of the laser-driven betatron radiation experiments are working at two major schemes: without [13] or with [16, 19] overlapped laser field. The oscillation amplitude of electrons can be significantly increased, when its betatron oscillation frequency resonates with the laser field [28]. Thus, the overlapped case is a more effective way to enhance betatron X-ray radiation. We have been studying laser-driven X-ray sources for decades with different scale and parameters of laser systems, and recently making a lot of progress in enhancing photon energy and yield of betatron radiation based on wakefield-accelerated electrons wiggling in laser fields.

In this article, we review the principle and properties of betatron radiation from accelerated electrons resonance with the laser field in Sect. 17.2. In Sect. 17.3, resonantly enhanced betatron hard X-ray radiation from ionization-injected electrons based on laser-wakefield acceleration (LWFA) is briefly reviewed [19]. In Sect. 17.4, we review a new method for generating betatron gamma ray radiation based on dual-laser pulse scheme [29]. Finally, in Sect. 17.5 is the conclusion and perspectives.

17.2 Betatron Radiation from Directed-Laser-Accelerated Electrons

Consider a linearly polarized plane wave with $E_{yL} = E_0 \cos(\varphi)$ and $B_{zL} = E_{Ly}/v_{ph}$, where E_0 is the amplitude of the laser field, $\varphi = k$x-ω_0t is the phase of the electric field, and $v_{ph} = \omega_0/k$ is the phase velocity of laser. The transverse self-generated electromagnetic field of the wakefield is written as E_{ys}, E_{zs}, $B_{ys,}$ and B_{zs}. Then, the transverse dynamics of an electron interaction with the linearly polarized laser field and the self-generated field can be described as

$$\frac{dp_y}{dt} = -eE_0\left(1 - \frac{v_x}{v_{ph}}\right)\cos(\varphi) - eE_{ys} + ev_x B_{zs} \tag{17.1}$$

When the velocity of the electron is close to c, dv_x/dt ~ 0 and d(1−v_x/v_{ph})/dt ~ 0, i.e., v_x is a slow variable compared to y and p_y. The phase seen by the relativistic electron can be written as $\varphi \approx \omega_L t = (1\text{-}v_x/v_{ph})\ \omega_0 t$, and ω_L is the Doppler-shifted laser frequency. Thus, (17.1) can be simplified as [30]:

$$\frac{d^2 p_y}{dt^2} + \omega_\beta^2 p_y = m_e c a_0 \omega_L^2 \sin(\omega_L t) \tag{17.2}$$

where $\omega_\beta = (e(\kappa_E + V_x\ \kappa_B)/(\gamma m_e))^{1/2}$ is the betatron frequency, $\kappa_E = \partial E_{ys}/\partial y = \partial E_{zs}/\partial z$, $\kappa_B = \partial B_{ys}/\partial z = -\partial B_{zs}/\partial y$, and $a_0 = eE_0/m_e\omega_0 c$. For resonant electrons with $\mathrm{v_x}$ ~ c, $\omega_\beta \approx \omega_p/(2\gamma)^{1/2}$. The oscillation amplitude will grow rapidly when betatron frequency is in resonance with the laser frequency ($\omega_\beta \approx \omega_L$). And, the electron transverse momentum can grow faster, when its betatron frequency varies by roughly twice the laser frequency ω_L.

As stated in the introduction, betatron radiation properties strictly depend on the electron dynamics, which in turn depends on laser and plasma parameters. Thus, radiation from direct-laser-accelerated electrons is quite different with that from only laser-wakefield-accelerated electrons. For direct-laser-accelerated electrons, the related radiation parameters scaling on laser intensity and plasma density as [30]

$$K \approx p_\beta/m_e c \approx 2^{1/4} a_0 (a_0 n_c/n_e)^{1/2}$$
$$E_c \propto (a_0)^3 (a_0 n_c/n_e)^{3/2},\ N_{\text{total}} \propto a_0 (a_0 n_c/n_e)^{1/2}, \theta = (n_e/(a_0 n_c))^{1/2}$$

While, for only laser-wakefield-accelerated electrons, the responding scaling laws are [9]: $K \propto a_0, E_c \propto a_0 (a_0 n_c/n_e)^{3/2}, N_{total} \propto a_0 (a_0 n_c/n_e)^{3/2}, \theta = (n_e/n_c)^{1/2}$. These scaling laws are obtained in bubble regime when matching condition is satisfied, i.e., $k_p w_0 = 2(a_0)^{1/2}$. Obviously, for a fixed value of $n_e/(a_0 n_c)$, radiation would be much stronger for direct-laser-accelerated case. The dynamics of electron is strongly dependent on the self-similar parameter of $n_e/(a_0 n_c)$ [30]. As wakefield-accelerated electrons interacting with a laser field, the strength parameter K would increase, resulting in radiation enhancement.

17.3 Resonantly Enhanced Betatron Radiation via Ionization Injection

Most of betatron radiation experiments were carried out based on self-injection mechanism. Due to the characteristics of self-injected electron acceleration scheme, both injection and acceleration are relatively long-term processes. Even if a relative high-energy laser is used to enable the injected electron to achieve resonance with the laser field [16], the optimal resonance timing is often missed due to the uncontrollability of the nonlinear process. Ionization injection [31] has a lower injection threshold and can improve the stability of the injection and acceleration processes, the injection condition is $\Delta\phi = \phi_{f-}\phi_i \leq -1$, where $\phi = e\Phi/mc^2$ is the normalized electrostatic potential of the plasma wave. Thus, Ionization injected electrons have a chance to be trapped before arriving at the tail of the wake, so that they can catch up with the laser pulse more quickly before pulse depletes dramatically.

We conducted experiments at the Key Laboratory of Laser Plasma of Shanghai Jiao Tong University in this way. In the experiment, we used 80 TW, 40 fs, $a_0 \approx 2.2$ laser pulse to interact with pure nitrogen (N_2), and obtained electron beam with energy peaked at 300 MeV and charge of 40 pC through ionization injection mechanism, along with collimated betatron hard X-ray beam, as shown in Fig. 17.1a, b. The betatron X-ray has a yield of 8×10^8/pulse, which is 10 times higher than that obtained by He case under similar laser conditions, and the number of photons with energy >110 keV reaches 10^8/pulse. The critical energy of the X-ray spectrum measured in nitrogen was $E_c = 75$ keV. The comparison of the current source with previous results [13, 16] was shown in Fig. 17.1c. Two-dimensional particle-in-cell (PIC) simulations were conducted with similar laser and gas parameters, and confirmed that enhanced betatron radiation results from "fast" ionization injection process, which shorten the time for injected electrons to catch up with laser pulse for simulating resonance enhancement of betatron oscillation, resulting in the oscillation amplitude and frequency greatly increasing, as shown in Fig. 17.1d, e. In addition, as shown in Fig. 17.1a, the elliptical profile of betatron beam for N_2 case implied that it has polarization along the direction of laser polarization [21]. This experiment proved a new generation method of stable, simple, and efficient excitation of high energy (polarized) betatron radiation source.

17.4 Betatron γ Ray Radiation Generation Based on Dual-Laser Pulses

As proved in [16, 19] betatron radiation can be enhanced via accelerated electrons resonating with laser field when they catch up with the tail of laser pulse. However, with only one single ultrashort laser pulse, through the long process of electron injection, acceleration, catching up with laser and resonance, the source's parameters

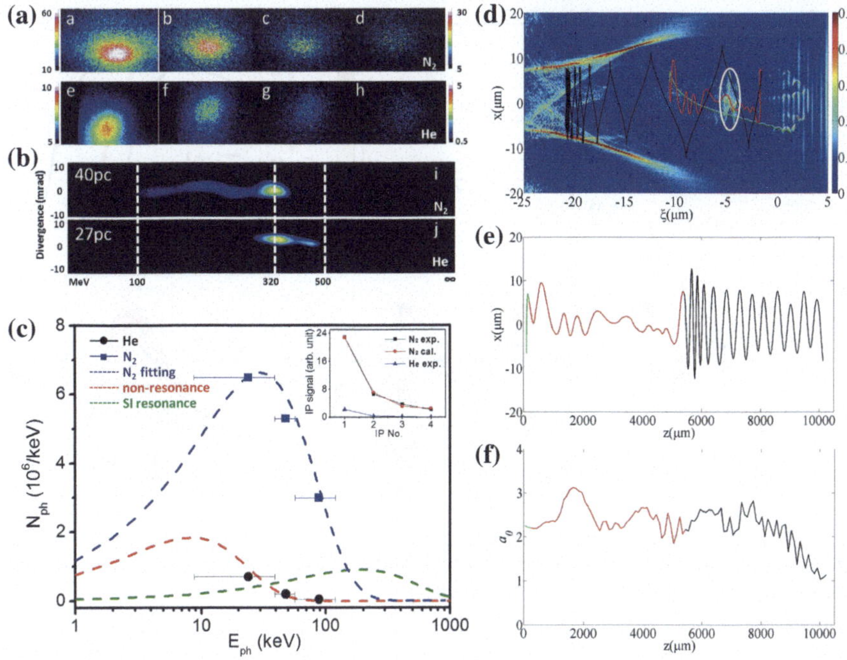

Fig. 17.1 **a**, **b** betatron X-ray signals and electron spectra with nitrogen and helium gases, respectively; **c** betatron X-ray spectrum analysis and comparison with [13, 16]. **d** a test electron in the reference frame of laser within the white dashed white circle in the electron density (normalized to its maximum density value) figure at t = 13.33 ps, $\xi = z - z_{max}$ (z_{max} is the location position of the peak laser field); **e** Trajectory of the same test electron; **f** The evolution of a_0

(photon number, energy, etc.) would be greatly reduced, and the effect of electron resonance cannot be fully exerted. First, electrons usually entering deceleration region of the wake when entering the laser field, resulting in a drop in critical energy E_c. Second, the laser intensity would soon or has already depleted seriously, as Fig. 17.1f showed, which limits the effectiveness of resonance. Moreover, with only one pulse, the acceleration and oscillation processes can't be controlled independently. While, one exception case is that the laser pulse length is long enough to cover the acceleration field. But, a longer laser pulse with a constant laser energy means smaller laser intensity, which results in a smaller electron energy gain [3]. Therefore, we propose a new method using dual ultrashort laser pulses that one laser drives wakefield acceleration and the other one wiggles accelerated electrons independently, which not only can sustain electrons located in the acceleration region of wakefield to ensure a high energy gain but also induce strong betatron oscillations in the wiggling laser field.

A series of PIC simulations were performed to compare the results of electron acceleration and betatron radiation for four different cases: case 1, only one pulse with 3 J, 30 fs; case 2, a longer pulse with 4.5 J, 60 fs; case 3, two pulses, the first (second) pulse is 3 J (1.5 J), 30 fs, and the delay of two pulses is 120 fs; case 4, two

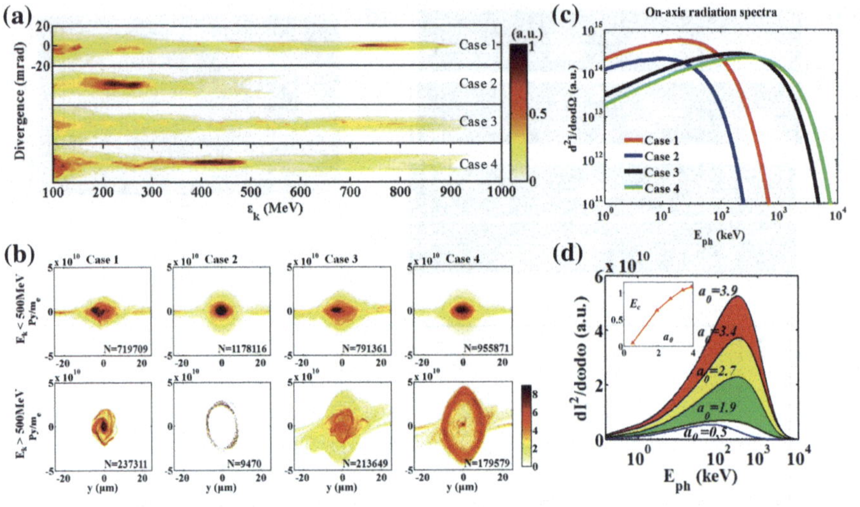

Fig. 17.2 **a** electron spectra for four different cases; **b** electron beam phase space of y-p_y for four different cases. **c** on-axis radiation spectra for four different cases; **d** on-axis radiation spectra and critical energy E_c for different a_0 of the second laser

pulses are same as case 3, but with delay of 60 fs. The spot sizes of two 800 nm p-polarized laser pulses were both $w_0 = 15$ μm. The neutral nitrogen longitudinal profile had a 100 μm up-ramp followed by a 6 mm long plateau with a uniform density of 6×10^{18} cm^{-3}, and the background pre-ionized plasma density was 3×10^{18} cm^{-3}. As shown in Fig. 17.2a, for a long pulse (case 2), the maximum electron energy is much lower; while, compared to case 1, for dual short pulse cases (case 3, 4), the maximum electron energy is slightly higher due to the contribution of direct laser acceleration (DLA) which is indicated by the bifurcation structure [32], and the high energy parts have larger divergence. The phase spaces (y, p_y) of electrons indicated the oscillation behavior, as shown in Fig. 17.2b, for all cases, the majority of electrons with energy below 500 MeV could oscillate stably in the bubble with amplitudes <5 μm. While, for electron energy above 500 MeV, the majority of electrons were located in an annulus with 7 μm < r < 10 μm for case 4, indicating that the interval of 60 fs was suitable for obtaining an electron beam with higher energy, a larger amplitude and stable oscillation in the bubble. Usually, high-energy electron beam with large oscillation amplitudes is more efficient to generate high-energy betatron radiation. As shown in Fig. 17.2c, the on-axis radiation spectra for these four cases were calculated. For case 4, the radiation spectrum has a maximum critical photon energy of E_c~820 keV. In contrast, case 1 corresponds to a critical photon energy of E_c ~ 75 keV. When accelerated electrons experience the second laser field, the wiggling field is one order of magnitude higher than that for the self-generated field (K ~ 11), and the equivalent strength parameter K ~ 187. The peak brightness of betatron radiation reaches 1.2×10^{23} ph/s/mrad^2/mm^2/0.1%BW at 1 MeV. Figure 17.2d shows betatron radiation spectra and critical energy E_c for different a_0 under optimal condition, which indicates that radiation can be further enhanced by using higher power laser. Our simulation results demonstrate that the proposed method is an efficient way to generate controllable and high yield betatron x/γ ray radiation, which relies on the intensity of the second laser pulse. In addition, the radiated photons should be polarized along the polarization direction of the second laser pulse, due to the oscillation of electrons that is mainly forced by its electric field.

17.5 Conclusion and Perspectives

In this paper, we briefly reviewed the principle and properties of betatron radiation from direct-laser-accelerated electrons, and summarized our recent progresses in betatron enhancement. First, by introducing ionization injection, the electrons can quickly catch up with the drive laser pulse and resonate with it, which leads to the increasing of the betatron oscillation amplitude. Second, high energy radiation was achieved by accelerated electrons wiggling in an additional laser field whose intensity is one order of magnitude higher than that for the self-generated transverse field of the bubble. All these progresses make laser-driven betatron radiation more efficient. In the near future, it's important to improve the stability, reproducibility and repetition rate of such X-ray sources for practical applications, such as computer tomography, X-ray absorption spectroscopy, etc.

References

1. T. Tajima, Laser electron accelerator. Phys. Rev. Lett. **43**, 175–267 (1979)
2. G.A. Mourou, Optics in the relativistic regime. Rev. Mod. Phys. **78**, 309–371 (2006)
3. E. Esarey, Physics of laser-driven plasma-based electron accelerators. Rev. Mod. Phys. **81**, 1229–1285 (2009)
4. A.J. Gonsalves, Petawatt laser guiding and electron beam acceleration to 8 GeV in a laser-heated capillary discharge waveguide. Phys. Rev. Lett. **122**, 084801 (2019)
5. W.T. Wang, High-brightness high-energy electron beams from a laser wakefield accelerator via energy chirp control. Phys. Rev. Lett. **117**, 124801 (2016)
6. J. Wenz, Dual-energy electron beams from a compact laser-driven accelerator. Nat. Phot. **13**, 263–269 (2019)
7. Y.F. Li, Generation of 20 kA electron beam from a laser wakefield accelerator. Phys. Plasmas **24**, 023108 (2017)
8. J.P. Couperus, Demonstration of a beam loaded nanocoulomb-class laser wakefield accelerator. Nat. Commun. **8**(1), 487 (2017)
9. S. Corde, Femtosecond x-rays from laser-plasma accelerators. Rev. Mod. Phys. **85**, 1–48 (2013)
10. A. Rousse, Production of a keV x-ray beam from synchrotron radiation in relativistic laser-plasma interaction. Phys. Rev. Lett. **93**, 135005 (2004)
11. F. Albert, Applications of laser wakefield accelerator-based light sources. Plasma Phys. Control. Fusion **58**, 103001 (2016)
12. E. Esarey, Synchrotron radiation from electron beams in plasma-focusing channels. Phys. Rev. E **65**, 056505 (2002)

13. S. Kneip et al., Bright spatially coherent synchrotron x-rays from a table-top source. Nat. Phys. **6**, 980–983 (2010)
14. K.T. Phuoc, Imaging electron trajectories in a laser-wakefield cavity using betatron x-ray radiation. Phys. Rev. Lett. **97**, 225002 (2006)
15. K.T. Phuoc, Demonstration of the ultrafast nature of laser produced betatron radiation. Phys. Plasmas **14**, 080701 (2007)
16. S. Cipiccia, Gamma-rays from harmonically resonant betatron oscillations in a plasma wake. Nat. Phys. **7**, 867 (2011)
17. L.M. Chen, Bright betatron x-ray radiation from a laser driven-clustering gas target. Sci. Rep. **3**, 1912 (2013)
18. W. Yan, Concurrence of monoenergetic electron beams and bright X-rays from an evolving laser-plasma bubble. Proc. Natl. Acad. Sci. U. S. A. **111**, 5825–5830 (2014)
19. K. Huang, Resonantly enhanced betatron hard x-rays from ionization injected electrons in a laser plasma accelerator. Sci. Rep. **6**, 27633 (2016)
20. M. Schnell, Optical control of hard x-ray polarization by electron injection in a laser wakefield accelerator. Nat. Commun. **4**, 2421 (2013)
21. A. Döpp, Stable femtosecond X-rays with tunable polarization from a laser-driven accelerator. Light: Sci. Appl. **6**, e17086 (2017)
22. J. Wenz, Quantitative X-ray phase-contrast microtomography from a compact laser-driven betatron source. Nat. Commun. **6**, 7568 (2015)
23. A. Döpp, Quick x-ray microtomography using a laser-driven betatron source. Optica **5**(2), 199–203 (2018)
24. J.M. Cole, High-resolution μCT of a mouse embryo using a compact laser-driven X-ray betatron source. Proc. Natl. Acad. Sci. U. S. A. **115**, 6335–6340 (2018)
25. M.Z. Mo, Measurements of ionization states in warm dense aluminum with betatron radiation. Phys. Rev. E **95**, 053208 (2017)
26. B. Mahieu, Probing warm dense matter using femtosecond X-ray absorption spectroscopy with a laser-produced betatron source. Nat. Commun. **9**, 3276 (2018)
27. J.C. Wood, Ultrafast imaging of laser driven shock waves using betatron x-rays from a laser wakefield accelerator. Sci. Rep. **8**, 11010 (2018)
28. A. Pukhov, Relativistic laser plasma interaction by multi-dimensional particle-in-cell simulations. Phys. Plasmas. **5**, 1880 (1998)
29. Feng Jie, Gamma ray emission from wakefield accelerated electrons wiggling in laser field. Sci. Rep. **9**, 2531 (2019)
30. T.W. Huang, Characteristics of betatron radiation from direct-laser-accelerated electrons. Phys. Rev. E **93**, 063203 (2016)
31. A. Pak, Injection and trapping of tunnel-ionized electrons into laser-produced wakes. Phys. Rev. Lett. **104**, 025003 (2010)
32. J.L. Shaw, Role of direct laser acceleration of electrons in a laser wakefield accelerator with ionization injection. Phys. Rev. Lett. **118**, 064801 (2010)

Chapter 18
LWFA-Driven Betatron Source for Plasma Physics Platform at ELI Beamlines

U. Chaulagain, K. Boháček, J. Vančura, M. Lamač, W. Yan, Y. Gu, M. Kozlová, K. Ta-Phuoc, S. A. Weber and J. Nejdl

Abstract In this paper, we report on the development of a plasma betatron X-ray source for active diagnostics of various plasma physics and high energy density physics experiments in the Plasma Physics Platform (P3) at ELI Beamlines. This X-ray source will deliver ultrashort X-ray pulses with photon energies ranging from a few keV up to 100's of keV with over 10^9 photons per pulse with a rep. rate up to 10 Hz. The driver laser for betatron source is a Ti:sapphire diode-pumped HAPLS laser system [1] that can deliver up to 30 J of energy in less than 30 femtosecond, at a maximum repetition rate of 10 Hz.

18.1 Introduction

In high energy density (HED) plasma, a very small volume of target (of few mm^3) is converted into extreme states of matter characterized by very high pressures (~Mbar) and very high temperatures (~10's eV) which can be found in many astrophysical processes such as evolving stars, accretion flows, core of the giant planets, supernova remnants, etc. HED state also generated terrestrially in dynamic processes for instance when asteroid impacts on earth, where matter are suddenly subjected to very high temperature and pressures resulting in a strong shock. Similar conditions

U. Chaulagain (✉) · K. Boháček · J. Vančura · M. Lamač · W. Yan · Y. Gu · M. Kozlová · S. A. Weber · J. Nejdl
ELI Beamlines, Institute of Physics ASCR, Prague, Czech Republic
e-mail: uddhab.chaulagain@eli-beams.eu

K. Boháček · J. Vančura
FNSPE, Czech Technical University in Prague, Prague, Czech Republic

M. Lamač
Charles University, Prague, Czech Republic

Y. Gu · M. Kozlová · J. Nejdl
Institute of Plasma Physics ASCR, Prague, Czech Republic

K. Ta-Phuoc
Laboratoire d'Optique Appliquée, ENSTA, CNRS UMR7639, École Polytechnique, Palaiseau, France

M. Kozlová and J. Nejdl (eds.), *X-Ray Lasers 2018*, Springer Proceedings in Physics 241, https://doi.org/10.1007/978-3-030-35453-4_18

are also found in the laboratory experiments related to nuclear fusion with inertial confinement. One key example of HED is a radiative shock (speed ~10's kms^{-1}), in which radiation couples with matter and modifies the dynamics of the shock structure [2–6]. The shock front and post-shock structure are extremely dense with the electron densities ranging from 10^{21} to 10^{26} cm^{-3}, and the shock front is very thin. The probing of such a complex structure is challenging. The behavior of materials in the extreme states of pressure and temperature has particular interests in condensed matter physics and material failure mechanisms [7]. The state-of-the-art large-scale high-power laser facilities are capable, today, of creating extreme high energy density plasmas relevant to astrophysical systems [8, 9]. Using a high-energy-pulsed laser, one can drive a strong shock into the material and create the extreme state of matter reaching high temperatures and densities at equilibrium relevant to planetary interiors. The commonly used method to measure the shock velocity in high-energy density matter is VISAR based on the optical laser. However, this method does not provide detailed information about the shock front and the post-shock structure as the visible laser cannot probe the dense plasma structure. The current X-ray sources used for imaging of the shock do not provide a high temporal resolution that is required to image these shocks and other fast phenomena as the radiation from synchrotrons is limited to a temporal resolution of ~100 ps [7].

Most of these laser-driven HED plasmas are in a nonequilibrium state; hence, they are extremely difficult to probe. Commonly used plasma probing tools in large-scale facilities are K-α X-ray sources which are isotropic (not highly directional) and not easily energy tunable. These X-ray sources have large source sizes and longer pulse durations, not achieving the necessary resolution to characterize HED plasmas. However, the short temporal and small spatial scales of laboratory-generated HED phenomena make their diagnostics challenging. To probe fast-moving and ultra-dense plasma, it requires a small source size, a highly directional and short pulse duration X-ray source.

Here we present a betatron X-ray source that is emitted by electrons accelerated and wiggled by the laser wakefield, and is very similar to synchrotron wiggler radiation, with the following features: a broadband continuous spectrum (1–100 keV), a narrow divergence (less than tens of mrad), a small source size (μm), a short pulse duration (~few fs) and an inherent synchronization with the driving pulse [10–13]. This source provides very high peak brightness, about six orders of magnitude higher than conventional X-ray backlighter sources because of its shorter pulse duration. The intrinsic temporal resolution of betatron radiation is about 10 fs, indicating that significantly faster processes could be probed without motion blurring and without compromising the spatial resolution. Also, due to the micron-scale source size, the LWFA-driven x-ray source possesses significant spatial coherence that also enables the phase-contrast imaging [14, 15].

18.2 LWFA-Driven Betatron Source at P3 Platform

The Plasma Physics Platform (P3) platform [16] is a flexible and versatile platform dedicated to fundamental research. P3 experimental chamber is made of aluminium and has a decagonal body of 4.5 m inner diameter and 3.4 m height with volume roughly 50 m^3. The P3 infrastructure is designed to handle multiple, synchronized laser beams in the same interaction chamber, covering a wide range of pulse lengths and energies. With its unique combination of laser beams, including the 10 PW, a high rep. rate PW laser and a kilo-Joule beam (ns pulses with temporal shaping), it will focus on wide range of laser-plasma physics including the ultra-intense laser-matter interaction, warm dense matter (WDM), collisionless shock waves, radiative shock waves, HED physics, and advanced plasma physics experiments. The betatron source developed for P3 serves as an ultrafast probe for various HED plasma experiments.

To produce a sufficient number of X-ray photons by betatron, we need to accelerate electrons to suitable energy and keep them oscillating within the wakefield on a longer path. The quality of the focal spot is crucial in identifying the achievable performance of the acceleration process. Well-matched laser pulse of above critical power and relativistic intensity self-focuses to the size w_0 in the order of plasma wavelength λ_p and guides itself along longer distance with spot size oscillating around the matched spot size. The optimal bubble regime is reached when $\lambda_p \sim 2c\tau$, where $c\tau$ is the length of the laser pulse, $c\tau \sim 10\ \mu m$ for $\tau = 30$ fs [17]. The focusing of the laser beam is thus a crucial topic for successful generation of betatron radiation. Although the optimal focal spot can be produced by using a parabolic mirror, its precise manufacturing can be complicated and expensive for longer focal lengths and larger sizes. Thus, especially for higher F number the spherical mirrors with normal incidence geometry can be used [18, 19], as the spherical aberration becomes rather small.

The driver laser for betatron source is a Ti:Sa diode-pumped HAPLS laser (L3 laser) that can deliver up to 30 J of energy in less than 30 femtosecond, at a maximum repetition rate of 10 Hz. The laser beam size is 215 mm ($1/e^2$). The experimental setup for electron acceleration and betatron generation is shown in Fig. 18.1. The L3 laser beam is delivered onto the focusing spherical mirror ($f = 5000$ mm), placed in a separate focusing chamber, by a mirror with hole, which the focused beam is sent through into a supersonic gas jet. The influence of the hole on the beam focus quality has been studied. It is seen that the simulation with the square beam focused by a spherical mirror provides comparable results as a theoretical limit of Airy disk focusing. The typical focal spot with this configuration is shown in Fig. 18.2. The choice of the focal length is defined to reach a laser strength parameter sufficiently high ($a_0 = 2$–4).

A supersonic gas jet or a gas cell will be used as a target. The choice of gas jet or gas cell will depend on the experiment. Various sizes of gas jets have been manufactured and characterized [20] for this purpose. The electron density of the plasma will be of the order of $10^{18}\ cm^{-3}$. The target length determines the length of acceleration, therefore the electron energy and, consequently, the energy of the radiation. After the gas jet, the electrons are deflected by a permanent dipole magnet, serving also

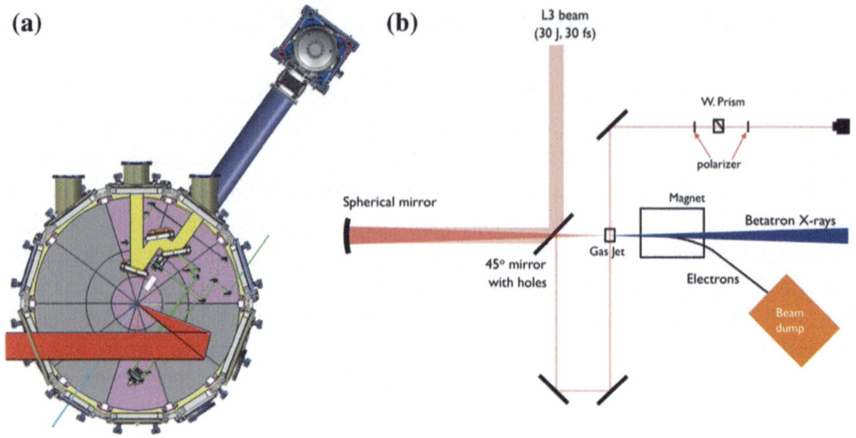

Fig. 18.1 **a** Focusing of L3 and L4 laser beams inside the P3 chamber. The L4 beam (10 PW) focuses at the target chamber center (TCC) of the chamber and creates HED plasmas which will be probed by the betatron X-ray beam (blue beam) **b** Experimental layout of the betatron generation via wakefield-accelerated electrons and its associate diagnostics, inside P3 chamber

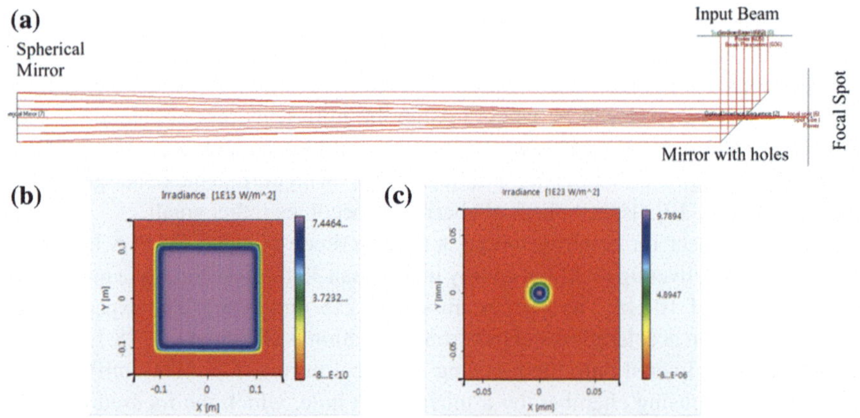

Fig. 18.2 **a** Optical simulation setup in Virtual Lab optical simulation package. The L3 laser beam incidents on the mirror with a hole, which reflects it to the spherical mirror and focuses though the mirror with hole. **b** Input square-shaped super-Gaussian field, **c** Typical focal spot and intensity profile at the focus considering an ideal square-shaped super-Gaussian beam (g = 20) showing the focal spot diameter is around 40 μm over $1/e^2$

as an electron spectrometer. A set of diagnostics will be used to characterize the laser-matter interaction, the produced electrons, and betatron radiation. Interaction diagnostics (Thomson scattering, shadowgraphy and/or interferometry) will provide information on the laser propagation in the gas.

Table 18.1 Expected electrons and betatron X-ray parameters in the experimental hall E3

Electron				Betatron X-ray		
Energy (GeV)	No. of electrons	Charge (pC)	Divergence (mrad)	Critical energy (keV)	Flux (/shot)	Divergence (mrad)
0.4–1.5	10^9	100	20	30	10^9	20

A 30 cm-long magnetic electron spectrometer, with a uniform magnetic field in the gap of 1 T, has been manufactured and tested with LWFA electron sources. The laser beam can be stopped by a thin metallic foil, thus only the betatron X-rays are reaching the X-ray diagnostics. Radiation diagnostics consists of single-photon counting mode CCD camera; low and high energy Ross filters pack are also designed and tested. These diagnostics will enable to assess the main characteristics of the generated radiation (divergence, flux, source size, spectrum). Since betatron sources produce radiation over a wide spectrum, several X-ray diagnostics will be implemented to cover a spectral range from a few keV up to 100 keV. All the necessary diagnostics for the plasma, electron, and X-ray have been prototyped and tested [21, 22].

Betatron source generates broadband synchrotron-like radiation with critical energy (median of spectral intensity) depending on the energy of the electrons as well as on the plasma density. For the first mode of operation, the L3 laser will be operated at 10 J, 30 fs and the rep. rate of 3.3 Hz. With the 10 J, 30 fs laser pulse, and the plasma electron density of 5×10^{18} cm^{-3}, the expected electron and X-ray parameters are reported in Table 18.1.

This X-ray source can be used for various applications. The broadband X-ray radiation in the range 1–20 keV will be used to probe matter via X-ray Absorption Spectroscopy (XAS). The LWFA-driven betatron radiation is a suitable candidate for laboratory-based time-resolved X-ray absorption applications due to their unique features: ultrashort pulse durations (~fs), a continuous spectrum from soft to hard X-ray energies, intrinsic all optical, and jitter-free synchronization with the pump pulse. In addition, a separate multidisciplinary, user-oriented hard X-ray Gammatron beamline is being designed in the experimental hall E2 at ELI beamlines [23].

18.3 Conclusion

The LWFA-driven betatron X-ray source could be an ideal backlighting source for different HED plasmas. In addition of a betatron X-ray source, it is also possible to use the ultrashort electron beam generated from LWFA in the range of MeV to GeV for electron radiography or in the production of hard X-ray source via inverse Compton scattering. These sources would greatly broaden the diagnostic capability of the P3 platform. This capability would be of a particular interest to the high energy density physics community and the study of warm dense matter (WDM), if used in conjunction with additional high-power/energy laser systems and betatron

X-ray source with femtosecond temporal resolution. Such a combined diagnostic capability would be unique among all laser facilities globally and provide a unique platform for the study of dense plasmas.

Acknowledgements Supported by the project Advanced research using high-intensity laser-produced photons and particles (CZ.02.1.01/0.0/0.0/16_019/0000789) from European Regional Development Fund (ADONIS). The results of the Project LQ1606 were obtained with the financial support of the Ministry of Education, Youth and Sports as part of targeted support from the National Programme of Sustainability II. The results of the Project LM2015083 were obtained with the financial support of the Ministry of Education, Youth and Sports within targeted support of Large infrastructures.

References

1. C.L. Haefner et al., High average power, diode pumped petawatt laser systems: a new generation of lasers enabling precision science and commercial applications, in *Research Using Extreme Light: Entering New Frontiers with Petawatt-Class Lasers III*, vol. 10241 (International Society for Optics and Photonics, 2017), p. 1024102
2. A.B. Reighard et al., Observation of collapsing radiative shocks in laboratory experiments. Phys. Plasmas **13**(8), 082901 (2006)
3. F. Suzuki-Vidal et al., Counterpropagating radiative shock experiments on the orion laser. Phys. Rev. Lett. **119**(5), 055001(2017)
4. U. Chaulagain et al., Structure of a laser-driven radiative shock. High Energy Density Phys. **17**, 106–113 (2015)
5. R.L. Singh et al., Experimental study of the interaction of two laser-driven radiative shocks at the PALS laser. High Energy Density Phys. **23**, 20–30 (2017)
6. T. Clayson et al., Counter-propagating radiative shock experiments on the Orion laser and the formation of radiative precursors. High Energy Density Phys. **23**, 60–72 (2017)
7. J.C. Wood, et al., Ultrafast imaging of laser driven shock waves using betatron x-rays from a laser wakefield accelerator (2018). https://arxiv.org/abs/1802.02119
8. D.D. Ryutov et al., Criteria for scaled laboratory simulations of astrophysical MHD phenomena. Astrophys. J. Suppl. Ser. **127**(2), 465 (2000)
9. B.A. Remington et al., A review of astrophysics experiments on intense lasers. Phys. Plasmas **7**(5), 1641–1652 (2000)
10. A. Rousse et al., Production of a keV X-ray beam from synchrotron radiation in relativistic laser-plasma interaction. Phys. Rev. Lett. **93**(13), 135005 (2004)
11. K. Ta Phuoc et al., Demonstration of the ultrafast nature of laser produced betatron radiation. Phys. Plasmas **14**(8), 080701 (2007)
12. O. Lundh et al., Few femtoseconds, few kiloampere electron bunch produced by a laser–plasma accelerator. Nat. Phys. **7**(3), 219 (2011)
13. V. Horny et al., Temporal profile of betatron radiation from laser-driven electron accelerators. Phys. Plasmas **24**, 063107 (2017)
14. S.N. Luo et al., Gas gun shock experiments with single-pulse x-ray phase contrast imaging and diffraction at the advanced photon source. Rev. Sci. Instrum. **83**(7), 073903 (2012)
15. U. Chaulagain et al., X-ray phase contrast imaging of biological samples using a betatron x-ray source generated in a laser wakefield accelerator, in *Laser Acceleration of Electrons, Protons, and Ions IV*, vol. 10240 (International Society for Optics and Photonics, 2017), p. 1024014
16. S. Weber et al., P3: An installation for high-energy density plasma physics and ultra-high intensity laser–matter interaction at ELI-Beamlines. Matter Radiat. Extrem. **2**(4), 149–176 (2017)

17. S. Corde et al., Femtosecond x rays from laser-plasma accelerators. Rev. Modern Phys. **85**(1) (2013)
18. V. Cipiccia et al., Gamma-rays from harmonically resonant betatron oscillations in a plasma wake. Nat. Phys. **7**(11), 867 (2011)
19. X. Wang et al., Quasi-monoenergetic laser-plasma acceleration of electrons to 2 GeV. Nat. Commun. **4**, 1988 (2013)
20. J. Nejdl et al., Imaging Michelson interferometer for a low-density gas jet characterization. Rev. Sci. Instr. **90**(6), 065107 (2019)
21. M. Kozlova et al., Laser-driven plasma-based incoherent x-ray sources at PALS and ELI beamlines, in *International Conference on X-ray Lasers*. Springer, Cham (2016), pp. 127–134
22. K. Bohacek et al., Stable electron beams from laser wakefield acceleration with few-terawatt driver using a supersonic air jet. Nucl. Instrum. Methods Phys. Res. Sect. A **883**, 24–28 (2018)
23. U. Chaulagain et al., ELI gammatron beamline: dawn of ultrafast hard X-ray science (in preparation)

Chapter 19
Imaging Objects with Coded Apertures, Utilising a Laser Wakefield X-Ray Source

M. P. Selwood, R. Heathcote and C. D. Murphy

Abstract Laser wakefield acceleration (LWFA) (Tajima and Dawson, P Rev Lett 43:267–270, 1979, [1]) is able to generate GeV electrons in millimetres of under dense plasma (Leemans et al., Nat Phys 2:696–699, 2006, [2]) due to the high fields which may be sustained. The electron beam is smaller than those from linear accelerators, and as such these bunches can be used to create hard X-ray imaging sources with micron resolution capability. The conventional pinhole imaging technique results in systems where improved resolution comes at the cost of flux reduction. At low flux, the resultant decrease in signal to noise ratio must be overcome by either a more powerful X-ray source or an increased integration time. Here, coded apertures are proposed as a viable imaging alternative for pinhole apertures without such adjustments. The imaging capabilities of coded apertures will be discussed and simulation results will be presented from ray-tracing algorithms.

19.1 Introduction

Laser wakefield acceleration (LWFA) is capable of generating femtosecond duration high energy X-ray sources with a micron-scale source size [3]. This allows for imaging systems capable of capturing still images from a rapidly evolving object with micron resolution. However, conventional pinhole imaging techniques are designed around the inverse correlation between the signal-to-noise ratio and resolution capability. To resolve 5 μm features on a typical 5×5 mm CCD chip, the open area ratio, and thus the throughput ratio, would be *ca.* 8×10^{-7}. This would necessitate a higher power source or long integration time; the former being costly, and the latter negating the ability to capture still images of a transient object.

M. P. Selwood (✉) · C. D. Murphy
Department of Physics, York Plasma Institute, University of York,
York YO10 5DQ, UK
e-mail: matthew.selwood@york.ac.uk

R. Heathcote
Central Laser Facility, STFC Rutherford Appleton Laboratory,
Didcot, Oxon OX11 0QX, UK

M. Kozlová and J. Nejdl (eds.), *X-Ray Lasers 2018*, Springer Proceedings
in Physics 241, https://doi.org/10.1007/978-3-030-35453-4_19

Instead a coded aperture can be used. These are an array of pinholes in specific arrangements, which each cast an image of the object onto the detector; the resultant signal from each overlaid image is dubbed the hologram. This needs to be decoded post-capture to form a likeness of the object. Coded apertures allow for a flux increase, as there are multiple pinholes, without resolution loss, as the individual pinholes are small in area. Using the MURA configuration [4] (discussed below), an open aperture area ratio of 0.5 is attainable. This is 6 orders of magnitude higher than the standard pinhole for the system described previously, and can still utilise the strobed nature of the LWFA source.

19.2 Image Decoding

A likeness of the object is decoded from the hologram via cross correlation algorithm; a quantitative measure of similarity between two signals as a function of lag [5]. Large values in the resultant array denote areas of high correlation between the two signals. For the coded aperture system, the two signals being compared are D and G; the hologram on the detector plane, and some decoding function, respectively, where the latter is constructed from the aperture design, A. These are decoded to form a likeness of the object, S, thus [4, 5]:

$$\begin{aligned}\hat{S}[i', j'] &= D[i, j] \star G[i, j] \\ &= \sum_{i'=0}^{i_{max}} \sum_{j'=0}^{j_{max}} D^*[i', j'] \cdot G[i' + i, j' + j]\end{aligned} \tag{19.1}$$

where D^* is the complex conjugate of D. Equation 19.1 can be visualised as a moving dot product; a window, G, is scanned across static array, D, and the dot product of the overlap is computed.

19.3 Coded Apertures

Under the assumption of intrinsic errors within the decoding process, combined with the fact that the detector plane is constructed such that $D = S \star A$, (19.1) becomes

$$\hat{S} = S \star (A \star G + N \star G) \tag{19.2}$$

where N is some noise term that is quasi-uniform across all spaces. In an idealised scenario, $\hat{S} = S$. For this to be reflected mathematically, G must be constructed such that $A \star G = \delta$, and $N \star G$ must equal 0. This is achieved by making G unimodular (values of ± 1) as a function of A, with a uniform density [6] to ensure that any superposition of G upon N within the moving dot product also has an expectation value of 0. This allows the technique to be uniform background subtractive, making it favourable for high noise environments. Furthermore, this may increase the SNR by a theorised factor of up to $\sqrt{\kappa}$, where κ is the number of perforations [7].

As G is constructed from A, thus A must also have a uniform perforation density. This will also reduce imaging bias between different regions of the field of view. Errors within $A \star G$ can arise if there is translational symmetry within the design, as the superposition would then show false correlation. Thus, the aperture design cannot hold translational symmetry, or have a constant perforation separation. This is achieved through similar techniques to pseudo-noise sequencing, with a prime number basis forming the aperture dimensions, p.

19.4 Modified Uniformly Redundant Arrays (MURA)

The MURA system is capable of creating numerous aperture designs, henceforth referred to as 'masks'. A is defined by a binary array, where a value of 1 denotes a perforated element, and 0 opaque sites. A and G have been created in accordance with the equations generated by Gottesman and Fenimore [4], and examples of different p basis apertures can be seen in their paper. An experimental advantage of these designs is the presence of column $A_0 = 0$, and row $A_0 = 1$, save for the first value. Upon cyclic permutation to bring the 0th row and column to the centre, a fiducial 'cross-hair' is created through the centre of the design for detector alignment.

One other the main advantage of coded aperture imaging is that they are uniform background subtractive, as G is unimodular by design. This is shown within Fig. 19.1; decoded point sources 19.1d,e are virtually identical and possess only a 11% SNR decrease, even though the latter hologram includes a uniform background level of 70%. In contrast, if a pinhole aperture were used, the SNR degradation would be 88%, equivalent to the visual difference between holograms 19.1a, b.

As (19.1) is a comparison of two arrays, it can be seen that a scaling factor will need to be applied to the size of G in order for the features to correlate with those on the detector. The scaling factor is applied to both the i and j coordinates, and is calculated thus:

$$M_T = \frac{A_{pix}}{D_{pix}} + \frac{v}{u} + 1 \tag{19.3}$$

where A_{pix} is the desired size of any single pinhole, v is the distance between the aperture and the detector, u is the distance from aperture to object, and the ratio between them is the magnification ratio, M_R. In an experimental system, the specific aperture design required is dictated by M_R, detector dimensions (D_{len}), and A_{pix} [8]:

$$p > A_{pix}\left[\frac{2D_{len}}{D_{pix}(M_R + 1)} - 1\right] \tag{19.4}$$

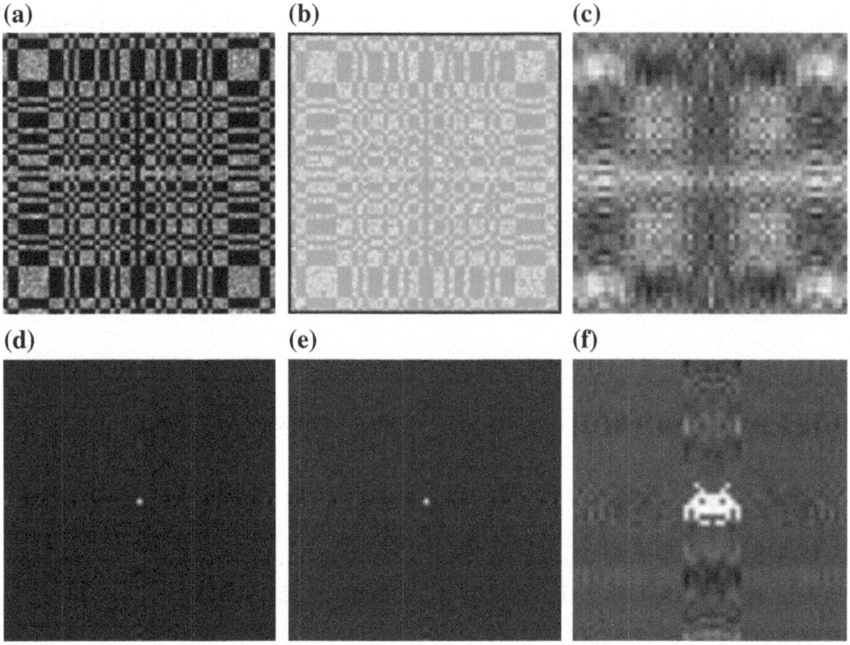

Fig. 19.1 Simulation results for a $p = 53$ aperture onto an idealised detector with $M_R = 1$, using 1×10^7 rays per point source. Holograms (**a**–**c**) were decoded to produce images of the point source, (**d**) and (**e**), and the extended object, (**f**). **b** is generated from (**a**), with the addition of a uniform background at 70% peak signal strength. **d** and **e** have an SNR of 60.1 and 53.4, respectively

where p is the smallest prime number that satisfies this criterion, and A_{pix} will be a function of the desired image resolution. As the MURA is a system of pinholes, it stands to reason that the resolution attainable would follow the same principles as that of a single pinhole [8]:

$$\frac{1}{R} = A_{pix}\left(\frac{1}{M_R} + 1\right) \tag{19.5}$$

where R is the resolution capability in the plane of the observed object, measured as the number of sources that can be resolved per unit length. As with the MURA's predecessor, the Uniformly Redundant Array (URA), the MURA possesses a theoretically quantifiable SNR [4, 9].

19.5 Simulating Holograms

A ray-tracing algorithm has been used to generate simulated holograms for decoding. A random ray, r is evaluated to see at what scalar extrapolation, τ, it crosses the plane of the aperture:

$$\tau = \frac{(\vec{P}_a - \vec{P}_s) \cdot \vec{n}}{\vec{r} \cdot \vec{n}} \tag{19.6}$$

where P_a is the position of the centre of the aperture, P_s is the position of the source, and n is the vector normal to the plane of the aperture, all in a constant reference frame. If τ is non-zero, the impact point of r on the aperture, P_i can be calculated:

$$\vec{P}_i = \vec{P}_s + \tau\, \vec{r} \tag{19.7}$$

If P_i is within the given aperture dimensions, it is compared against the binary sequence of A to test if the impact is a perforation site. All rays that are enabled to pass the aperture plane undergo (19.6) and (19.7) once more, using the position of the detector, P_d and its dimensions instead of the aperture. The number of rays impacting on each pixel of the detector can then be summed, and D built.

19.6 Conclusion

In conclusion, coded apertures are a viable imaging technique to fully utilise the improved resolution capabilities of laser wakefield acceleration X-ray sources. They are able to conserve resolution capabilities of conventional pinhole methods whilst drastically increasing the SNR. Their properties have been presented, and a ray-tracing methodology for hologram construction.

Acknowledgements This work was supported by the Engineering and Physical Sciences Research Council grants [EP/L01663X/1] and [EP/R512230/1], Scitech Precision, and the Science and Technology Facilities Council.

References

1. T. Tajima, J.M. Dawson, P. Rev. Lett. **43**, 267–270 (1979)
2. W.P. Leemans et al., Nat. Phys. **2**, 696–699 (2006)
3. S.P.D. Mangles et al., Nature **431**, 535–538 (2004); C.G.R. Geddes et al., Nature **431**, 538–541 (2004); J. Faure, et al., Nature **431**, 541–544 (2004)
4. S.R. Gottesman, E.E. Fenimore, Appl. Opt. **28**, 4344–4352 (1989)
5. A. Papoulis, *The Fourier Integral and Its Applications* (McGraw-Hill, New York, 1962)
6. E.E. Fenimore, T.M. Cannon, Appl. Opt. **17**, 337–347 (1978)
7. R.H. Dicke, Ap. J. **153**, L101–L106 (1968)

8. R. Heathcote et al., *Proceedings of SPIE Optical Engineering + Applications, San Diego*, vol 10763 (San Diego, CA, 2018), pp. 107630U-1–107630U-8
9. E.E. Fenimore, Appl. Opt. **22**, 3562–3570 (1978)

Chapter 20
Progresses of a Hard X-Ray Split and Delay Line Unit for the MID Station at the European XFEL

W. Lu, B. Friedrich, T. Noll, K. Zhou, J. Hallmann, G. Ansaldi, T. Roth, S. Serkez, G. Geloni, A. Madsen and S. Eisebitt

Abstract The Materials Imaging and Dynamics (MID) instrument at European XFEL aims at the investigation of nanosized structures and nanoscale dynamics using coherent X-ray radiation. A hard X-ray Split and Delay Line unit (SDL) is under manufacture for the MID station. The SDL is optimized to operate in a photon energy range from 5 to 10 keV and provides pairs of jitter-free X-ray pulses with a variable time delay ranging from −10 to 800 ps. With the powerful tunable and synchronized optical laser system at the MID station, not only X-ray pump-probe experiments but also X-ray probe—optical pump—X-ray probe (XOX) and optical pump—X-ray probe—X-ray probe (OXX) experiments are enabled by the SDL. This device will allow the study of ultrafast dynamics using experimental techniques, e.g. time-resolved X-ray Photon Correlation Spectroscopy (XPCS), Speckle Visibility Spectroscopy (SVS), ultrafast X-ray tomography and temporally and spatially resolved X-ray holography. We present here the current status of the mechanical manufacture of the device. Throughput simulations using diamond crystals as X-ray optics have been performed for the optical splitting scheme in SASE configuration and the results are presented.

K. Zhou conducted this research while being a visiting scholar at European XFEL.

W. Lu (✉) · J. Hallmann · G. Ansaldi · S. Serkez · G. Geloni · A. Madsen
European X-Ray Free-Electron Laser Facility, Holzkoppel 4, 22869 Schenefeld, Germany
e-mail: Wei.lu@xfel.eu

B. Friedrich · T. Noll · S. Eisebitt
Max Born Institute, Max-Born-Strasse 2A, 12489 Berlin, Germany

K. Zhou
Shanghai Institute of Applied Physics, Chinese Academy of Sciences, 201800 Shanghai, China

University of Chinese Academy of Sciences, 100049 Beijing, China

T. Roth
ESRF – The European Synchrotron, 71 Avenue des Martyrs, 38000 Grenoble, France

S. Eisebitt
Insitut für Optik und Atomare Physik, Technische Universität Berlin, 10623 Berlin, Germany

M. Kozlová and J. Nejdl (eds.), *X-Ray Lasers 2018*, Springer Proceedings in Physics 241, https://doi.org/10.1007/978-3-030-35453-4_20

20.1 Introduction

The intense, ultra short and coherent X-ray pulses provided by X-ray Free-Electron Lasers (XFELs) already opened up areas of research with X-rays that were previously inaccessible. An X-ray Split-and-Delay Line (SDL) unit enables X-ray pump-probe capability to the scientific instruments, leading to further improvement of the potential applications of XFELs. These devices will allow studies of ultrafast dynamics using various two-pulse X-ray experimental techniques. In addition, with powerful tunable and synchronized optical laser systems, not only X-ray pump-probe experiments but also X-ray probe—optical pump—X-ray probe (XOX) and optical pump—X-ray probe—X-ray probe (OXX) experiments are enabled.

The Materials Imaging and Dynamics (MID) instrument at the European XFEL facility [1, 2] aims at the investigation of nanoscale structure and dynamics by X-ray scattering and imaging. A hard X-ray SDL unit based on Bragg diffraction is under manufacture. It will be permanently installed at the MID station and available for user experiments as a default option of the instrument. We present here the concept of the device and the current status of mechanical manufacture. Moreover, instead of using Si crystals, we propose to utilize diamond crystals as an intensity beam splitter and merger for the SDL. Throughput simulations have been performed in SASE configuration and the results are presented.

20.2 Design and Manufacture Status

The European XFEL facility provides X-ray pulses separated by 220 ns (4.5 MHz) in 0.6 ms long bunch trains arriving with a repetition rate of 10 Hz [1]. Special operation modes [3] may permit the pulse spacing within the trains to be reduced to ~770 ps (defined by the accelerator RF frequency of 1.3 GHz) for a few pulses per train. The shorter time separation between individual pulses cannot be provided by the accelerator. The SDL unit at MID is designed to provide delays throughout the femtosecond and picosecond range up to 800 ps, allowing to cover this time-domain inaccessible by double pulses from the accelerator.

The conceptual design of the SDL is presented in Fig. 20.1. The incoming FEL pulse is separated into two parts by a beam splitter and the split pulses take two different trajectories (upper and lower branch), then they recombine at the merger and reach the sample on the same spot. By changing the path length of the upper branch, the difference in arrival times (Δt) between the two pulses can be varied from 0 to 800 ps with a few fs precisions for photons with energy between 5 and 10 keV. Two channel-cut crystals are employed in the lower branch to extend the optical path and thus allow for negative delay time between the two pulses. Two pairs of splitter and merger stages have been implemented in the design for intensity splitting [4] and geometrical splitting [5]. Moreover, the beam merger can be adjusted such

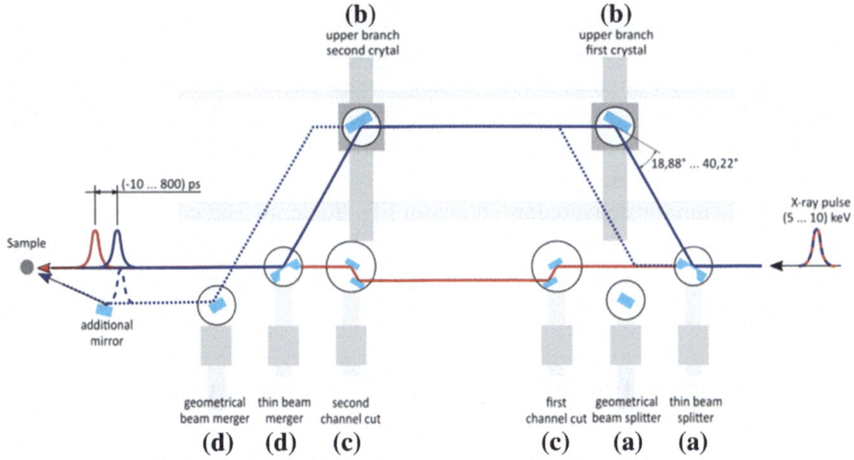

Fig. 20.1 Schematic layout of the split and delay line (SDL). **a** beam splitters; **b** upper branch crystals; **c** channel cuts; **d** beam merger

that either a collinear or a parallel, the non-collinear mode is provided. In the non-collinear mode, the two diffraction patterns resulting from interaction with a sample will be spatially separated on the detector [6].

In order to enable window-less operations of the beamline, the SDL will be operated in an ultra-high vacuum (UHV) environment. Figure 20.2a presents the view of the mechanical design of the SDL. Most of the components are customized products. The vacuum vessel, supported by a massive granite, is limited to 2 m length due to space constraints in the optic hutch. The high-stiffness L-shaped optical bench inside the vessel acts as the supporting structure for the precision mechanics and other components. The splitter cages (1), (2), channel-cut cages (3), (4) and merger cages (5), (6) in the lower branch are mounted on the vertical side of the L bench.

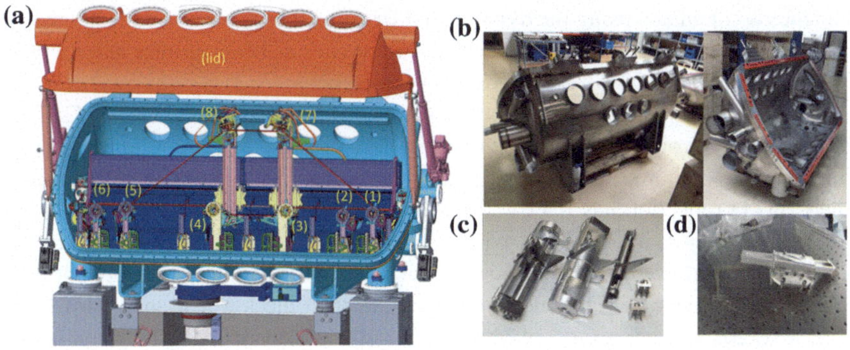

Fig. 20.2 **a** Mechanical design of the SDL. **b** Photo of the half-finished vessel. **c** In-vacuum 3 axis laser interferometer assemblies. **d** Beam diagnostic detector arm

The two upper branch crystal cages (7), (8) can move in X and Y direction along the rails attached on the L bench to provide the delays. Several on- and off-line diagnostic systems, e.g., alignment laser system, beam intensity diagnostic and, laser interferometry system are also implemented inside the vessel. The bottom part of the vacuum vessel has been manufactured, as shown in Fig. 20.2b. The complete vessel and the support granite are to be completed shortly. All inner mechanics are currently under manufacture or delivered as shown in Fig. 20.2c, d. The complete device is expected to be installed in the middle of 2019, followed by thorough commissioning with the beam.

20.3 Throughput Simulations with Diamond Crystals

The performances when using thin and thick Si crystals as either intensity or geometrical splitters, respectively, have been analyzed in our previous studies [7, 8]. Both of these concepts have been demonstrated in experiments [4, 5, 9]. Owing to easy alignment and reliable operation, devices based on geometrical splitting have recently been deployed at LCLS and SACLA [10]. In contrast, given the current technical difficulties associated with the fabrication and stable operation of thin, perfect silicon crystals [11], the intensity splitting concept, ideally in a 1:1 intensity splitting ratio, could so far not be widely used. Diamond is a superior material for many X-ray applications, as it has high transparency for hard X-rays, unrivaled radiation hardness and excellent thermal properties. Thin perfect diamond crystals with a few tens of micro-thickness are available [12]. Therefore, we propose using diamond thin crystals as a beam splitter and beam merger for the intensity splitting. We have compared the performance of thin C (111) and C (220) crystals via simulations, which provide promising results as shown below.

The throughput simulations use the same framework as our previous studies [8]: 20 fs FEL pulses with 8 keV photon energy are simulated with Genesis 1.3 [13] for European XFEL's SASE configuration. The photon transmission in both SDL branches is calculated using the dynamical diffraction theory based on the geometry sketched in Fig. 20.1 with C (111) or C (220) crystals as beam splitter and merger. The remaining crystals still employ Si (220) reflections.

As examples, Fig. 20.3a, b presents spectral data using 10 μm C (111) and C (220) thin crystals with single-color mode (at 7991.47 eV) in the SASE configuration [8]. Figure 20.3c, d shows the zoom-in of output "spike" in (a) and (b), respectively. The C (111) reflection has 0.5 eV Darwin width at this photon energy, similar to Si (220). As shown in Fig. 20.3c with a logarithmic scale, 99.9% of energy in the chosen spectral range is reflected the upper branch, which results in the upper branch having three orders of magnitude higher output than the lower branch. However, using C (220) instead, a ratio of almost 1:1 pulse energy splitting between upper and the lower branch has been observed. Since the Darwin width of C (220) is only 0.24 eV, the lower branch spectral width is defined by the Si (220) crystal behind the Diamond beam splitter. From Fig. 20.3d, one can easily see the 0.6 eV width lower

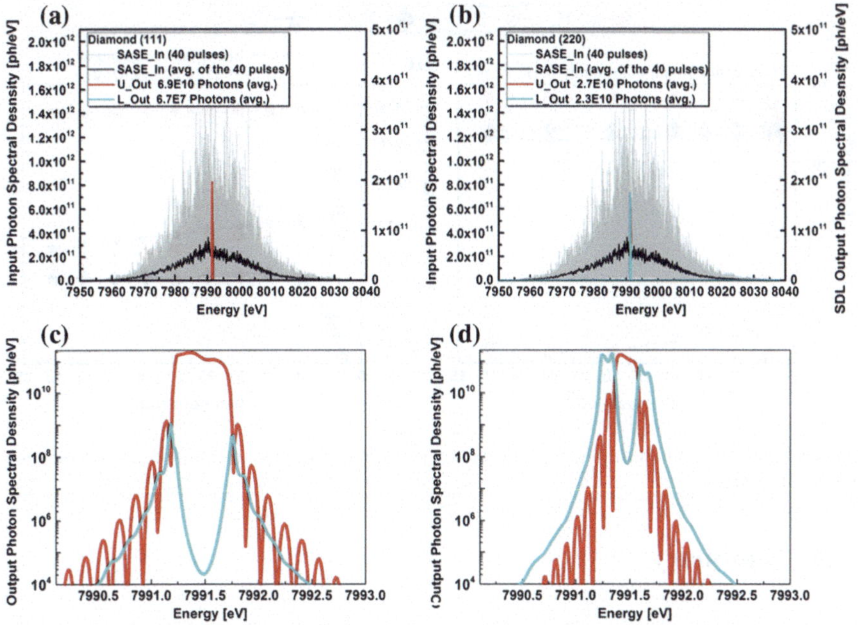

Fig. 20.3 Simulations for the spectral properties and throughput of the SDL device. The input spectra of the SDL are presented in grey (individual) and black (average), the outputs of the two branches are presented in red and cyan for the upper (U) and lower (L) branch of the SDL. The simulations have been performed using SASE radiation input with **a** C (111) and **b** C (220) splitter at single-color mode. **c** and **d** show the zoom-in of output "spike" in (**a**) and (**b**), respectively

branch output with a hole of 0.24 eV in the middle, which has been taken out to the upper branch by the Bragg reflection. The integration of the two spectra results in a similar pulse energy value. In the two-color operation mode of the SDL [8], the upper and lower branches pick up two different spectral components from the input spectrum. The Darwin width difference between the crystals does not play a critical role.

In Fig. 20.4, the output photon numbers are plotted as a function of crystal thickness. As can been seen in the single-color mode (a), the output changes dramatically with thickness when using C (111), but is not affected much in the case of using C (220). As explained above, the reason for this behaviour lies in the spectral properties. In the two-color mode (b), upper branch signals increase and saturate when the crystal thickness reaches beyond 10 μm. Lower branch signals decrease linearly with thickness due to absorption. We note that the C (220) case generates a better signal ratio between the upper and lower branch. We conclude that C (220) has superior performance as an intensity beam splitter reflection in the configuration studied.

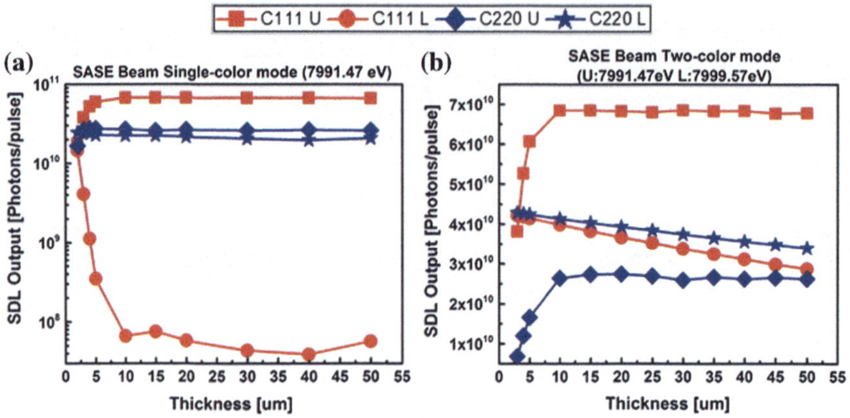

Fig. 20.4 Output photon numbers as a function of crystal thickness at **a** single-color mode and **b** two-color mode

20.4 Summary

In summary, we have presented the design of the SDL unit at the MID station of European XFEL and the current progress of the mechanical manufacturing of the device. We propose using thin diamond crystals as intensity beam splitters for the SDL and performed throughput simulations based on this concept. A comparison between using C (111) and C (220) is shown and we conclude that C (220) is better suited for this purpose.

Acknowledgements Funding via BMBF Verbundforschung under contracts 05K13KT4 and 05K16BC1 is gratefully acknowledged.

References

1. M. Altarelli et al., XFEL Technical Design Report, http://xfel.desy.de/technical_information/tdr/tdr/
2. A. Madsen et al., Technical Design Report of the MID instrument, http://pubdb.xfel.eu/record/154260
3. O. Grimm et al., Proceedings of EPAC 2006, Edinburgh, Scotland; A. Marinelli et al., Nat. Commun. **6**, 6369 (2015)
4. T. Osaka et al., Opt. Express **21**, 2823 (2013)
5. S. Roling et al., Proc. SPIE **8504**, 850407 (2012)
6. J.J. van Thor et al., Struct. Dyn. **2**(1), 014102 (2015)
7. W. Lu et al., AIP Conf. Proc. **1741**, 030010 (2016)
8. W. Lu et al., Rev. Sci. Instrum. **89**, 063121 (2018)
9. T. Osaka et al., IUCrJ **4**, 728 (2017)

10. Private communication with the XFEL facility users
11. Y. Feng et al., J. Phys. Conf. Ser. **425**, 052002 (2013)
12. U. Boesenberg et al., Opt. Express **25**, 279284 (2017)
13. S. Reiche, Nucl. Instr. Meth. Phys. Res. A **429**, 243 (1999)

Chapter 21
Modeling Diamond Irradiated with a European XFEL Pulse

N. Medvedev

Abstract Diamond irradiated with an intense femtosecond X-ray free-electron laser (XFEL) pulse is studied theoretically. The modeling is performed with the help of recently developed hybrid approach XTANT. The model includes processes of initial X-ray photon absorption resulting in photoelectron excitation, secondary electron cascades, Auger decays of K-shell holes, energy exchange between electron and atoms (electron-phonon coupling), nonthermal evolution of the atomic potential energy surface, atomic dynamics, and phase transitions. The results show that time-resolved graphitization of diamond irradiated with 24-keV photon energy pulses, as planned to be delivered at the European XFEL, takes slightly longer in comparison to the case of soft-X-ray irradiation.

21.1 Introduction

It is anticipated that the recently commissioned European XFEL, an X-ray free-electron laser, will be able to deliver ultrashort pulses (~30 fs FWHM), with up to 24–25 keV photon energy, and of sufficiently high intensity to induce phase transitions in irradiated targets by a single shot [1]. Such photon energies instigate a sequence of processes in irradiated solids, starting with core-shell photoabsorption releasing a highly energetic photoelectron and leaving a core hole. The primary electron is then performing secondary inelastic scattering, ionizing the next generations of electrons. Simultaneously, the core hole decays predominantly via Auger decay, emitting an Auger electron also capable of ionizing secondaries. These electronic scattering processes develop into an electron cascade at sub-picosecond timescales, increasing the number of excited electrons, see e.g. [2]. In case of elastic scattering, electrons exchange energy with the atomic lattice. This process is usually referred

N. Medvedev (✉)
Institute of Physics, Czech Academy of Sciences, Na Slovance 2, Prague 8, 18221 Prague, Czechia
e-mail: medvedevn@fzu.cz

Institute of Plasma Physics, Czech Academy of Sciences, Za Slovankou 3, Prague 8, 18200 Prague, Czechia

M. Kozlová and J. Nejdl (eds.), *X-Ray Lasers 2018*, Springer Proceedings in Physics 241, https://doi.org/10.1007/978-3-030-35453-4_21

to as electron-ion (or electron-phonon, in case if phononic approximation holds [3]) coupling, which heats up the atomic system [3, 4].

In covalent semiconductors and insulators, there is an additional mechanism by which an electronic system significantly affects atomic behavior: a nonthermal one [5]. Excitation of electrons across the bandgap from bonding to antibonding states leads to a modification of an interatomic potential. Atoms experiencing new forces may depart from their equilibrium positions and end up in a new phase. This process takes place independently of electron-ion coupling and without significant exchange of the kinetic energy between the electronic and atomic system, as a pure effect of the potential energy change. Hence, it is called a nonthermal phase transition [5].

At these femto- to pico-second timescales, electronic ensemble is out of equilibrium, that is, an electronic distribution function does not obey Fermi–Dirac distribution [2, 6]. In the case of irradiation with an XFEL pulse, the above-mentioned electron cascades are the main driving force of nonequilibrium.

It was demonstrated in previous works that diamond exposed to soft-X-ray FEL irradiation undergoes an ultrafast graphitization: a nonthermal solid–solid phase transition occurring within $\sim$150 fs [7]. In this paper, it is investigated what effect the electronic cascades induced by a hard X-ray have on the graphitization process.

21.2 Model

To model all the important processes taking place in XFEL irradiated diamond, the hybrid code X-ray-induced Thermal And Nonthermal Transitions (XTANT) is employed [2]. All the details of the code can be found in the review [2]; here, only a brief overview of the hybrid approach is given.

X-ray photoabsorption, high-energy electrons cascades as well as elastic scattering on target atoms, and Auger decays are described with the help of a classical Monte Carlo model. It allows to trace a nonequilibrium evolution of a highly excited fraction of electrons. These relaxing electrons exchange their energy and particles with the low-energy fraction of electrons (those with energies below 10 eV counted from the bottom of the conduction band), which are described with rate equations. Low-energy electrons populating the valence and conduction band also exchange their energy with target atoms, calculated with help of the Boltzmann collision integral. The probability of scattering entering the collision integral is calculated within the dynamical coupling approach beyond the Fermi golden rule [3].

Atomic motion is traced within the tight-binding molecular dynamics simulation. The transferable tight-binding Hamiltonian parameters for carbon are taken from [8]. An evolving electron distribution function, as described above, enters the calculations of interatomic forces, which naturally accounts for nonthermal effects and allows for tracing nonthermal graphitization [2, 7].

21.3 Results

XTANT simulations were performed for diamond irradiated with XFEL pulse with expected parameters of the European XFEL: 24 keV photon energy, 30 fs FWHM pulse duration, and 0.85 eV/atom absorbed dose. The snapshots of atomic positions are shown in Fig. 21.1, together with calculated powder diffraction patterns. The results clearly demonstrate graphitization taking place within approximately 200 fs since the maximum of XFEL pulse.

In the case of X-ray pulse, the graphitization takes longer than in the case of a soft-X-ray irradiation reported earlier (~150 fs [7]). This is a consequence of the longer lasting electron cascades [9]. Indeed, Fig. 21.2 illustrates the increase of the number of low-energy electrons excited into the conduction band. One can see that the number of electrons is increasing up to ~130–150 fs. Then, the increase slows down, before experiencing a second sharp rise.

The first increase is attributed to the nonequilibrium electron cascading, lasting significantly longer for a hard X-ray FEL. The second increase between ~150–190 fs is a result of the collapse of the bandgap, as can be seen in Fig. 21.3 [9].

A slow decrease of the density of conduction band electrons afterward is due to cooling down via electron-phonon coupling, which slowly returns electrons from the conduction to valence band. This cooling correspondingly heats up the atomic system, however, it takes place mainly *after* the phase transition is already completed. This supports the notion that the induced graphitization is of a purely nonthermal nature.

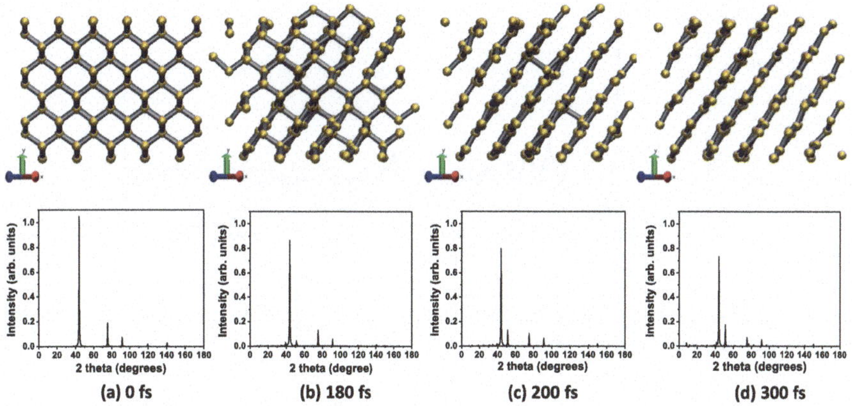

Fig. 21.1 (Top panel) Atomic snapshots of diamond irradiated with 24-keV photon energy, 30 fs FWHM laser pulse with an absorbed dose of 0.85 eV/atom. (Bottom panel) Corresponding powder diffraction patterns at 1.54 Å probe wavelength

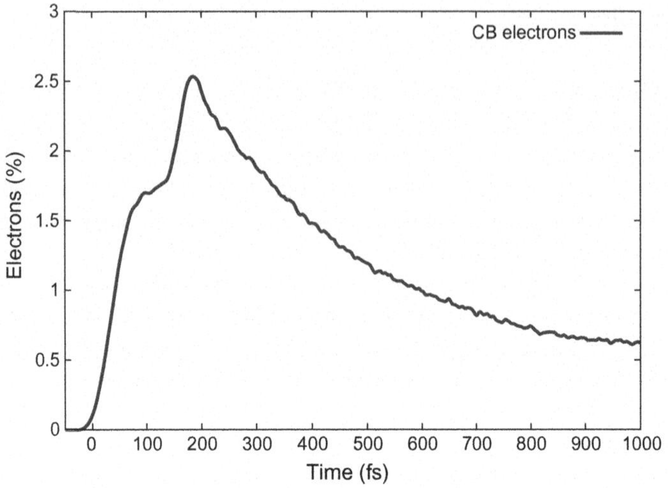

Fig. 21.2 Calculated fraction of conduction band electrons in diamond irradiated with 24-keV photon energy, 30 fs FWHM laser pulse corresponding to absorbed dose of 0.85 eV/atom

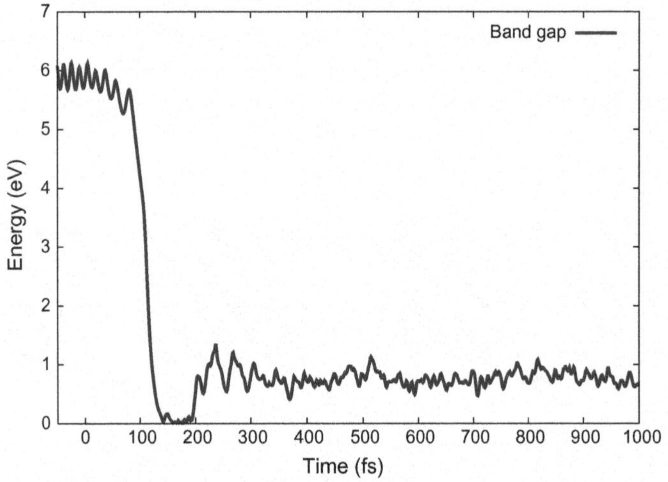

Fig. 21.3 Calculated band gap in diamond irradiated with 24-keV photon energy, 30 fs FWHM laser pulse corresponding to absorbed dose of 0.85 eV/atom

21.4 Conclusions

This paper presented the results of XTANT modeling of XFEL irradiation of diamond. It was found that nonthermal graphitization takes place within ~200 fs. It takes longer than the corresponding process under soft-X-ray irradiation due to longer nonequilibrium cascading time, delaying the energy deposition into the low-energy electronic fraction populating the valence and the bottom of the conduction band. After the energy is delivered there, the ensuing nonthermal solid–solid phase transition proceeds in essentially the same manner as in the case of visible or soft-X-ray femtosecond irradiation.

Acknowledgements The author thanks Libor Juha, Zheng Li, Vladimir Lipp, Victor Tkachenko, and Beata Ziaja for valuable discussions. Partial financial support from the Czech Ministry of Education (Grants LTT17015 and LM2015083) is acknowledged.

References

1. M. ltarelli, R. Brinkmann, M. Chergui, W. Decking, B. Obson, S. Düsterer, G. Grübel, W. Graeff, H. Graafsma, J. Hajdu, J. Marangos, J. Pflüger, H. Redlin, D. Riley, I. Robinson, J. Rossbach, A. Schwarz, K. Tiedtke, T. Tschentscher, I. Vartaniants, H. Wabnitz, H. Weise, R. Wichmann, K. Witte, A. Wolf, M. Wulff (eds.), *The European X-Ray Free-Electron Laser Technical design report*. (DESY XFEL Project Group European XFEL Project Team Deutsches Elektronen-Synchrotron Member of the Helmholtz Association, Hamburg, Germany, 2007)
2. N. Medvedev, V. Tkachenko, V. Lipp, Z. Li, B. Ziaja, Various damage mechanisms in carbon and silicon materials under femtosecond X-ray irradiation. **4**, 1–3 (2018)
3. N. Medvedev, Z. Li, V. Tkachenko, B. Ziaja, Electron-ion coupling in semiconductors beyond Fermi's golden rule. Phys. Rev. B **95**(1), 014309 (2017)
4. B. Rethfeld, D. Ivanov, M. Garcia, S. Anisimov, Modelling ultrafast laser ablation. J. Phys. D Appl. Phys. **50**(19), 193001 (2017)
5. S.K. Sundaram, E. Mazur, Inducing and probing non-thermal transitions in semiconductors using femtosecond laser pulses. Nat. Mater. **1**(4), 217–224 (2002)
6. B. Rethfeld, A. Kaiser, M. Vicanek, G. Simon, Ultrafast dynamics of nonequilibrium electrons in metals under femtosecond laser irradiation. Phys. Rev. B **65**(21), 214303 (2002)
7. F. Tavella, H. Höppner, V. Tkachenko, N. Medvedev, F. Capotondi, T. Golz, Y. Kai, M. Manfredda, E. Pedersoli, M.J. Prandolini, N. Stojanovic, T. Tanikawa, U. Teubner, S. Toleikis, B. Ziaja, Soft x-ray induced femtosecond solid-to-solid phase transition. High Energy Density Phys. **24**, 22–27 (2017)
8. C.H. Xu, C.Z. Wang, C.T. Chan, K.M. Ho, A transferable tight-binding potential for carbon. J. Phys. Condens. Matter **4**(28), 6047–6054 (1992)
9. N. Medvedev, Femtosecond X-ray induced electron kinetics in dielectrics: application for FEL-pulse-duration monitor. Appl. Phys. B **118**(3), 417–429 (2015)

Part IV
X-Ray Optics, Imaging and Metrology

Chapter 22
Time-Resolved Near-Edge X-Ray Absorption Spectroscopy in the Water Window and Beyond Using Laboratory-Based Laser Plasma Sources

H. Stiel, D. Engel, L. Glöggler, A. Jonas, I. Mantouvalou and J. Tümmler

Abstract In this contribution, we discuss the experimental requirements for time-resolved near-edge X-ray absorption fine structure (NEXAFS) spectroscopy in the soft X-ray range between 100 and 1500 eV in the laboratory. This photon energy range covers the so-called water window as well as the L-edges of transition metals and the M-edges of lanthanoids. We present a setup using a laser-produced plasma source and reflection zone plate optics. The setup allows the registration of NEXAFS spectra with a spectral resolution $\lambda/\Delta\lambda$ up to 1000 and time resolution ranging from few nanoseconds to picoseconds.

22.1 Introduction

Time-resolved X-ray absorption spectroscopy is a powerful method to gain both structural and functional information from molecules or functionalized materials [1, 2]. Whereas near-edge X-ray absorption fine structure spectroscopy (NEXAFS) mainly delivers information about electronic properties of unoccupied molecular orbitals extended X-ray absorption fine structure (EXAFS) spectroscopy is a powerful tool for exploring local structure around selected elements [2]. Combining the high sensitivity of both methods with a pulsed excitation of the sample dynamical changes of the molecular and electronic structure as well as energy or charge transfer processes can be explored [3]. Due to its relevance for life science and technology organic molecules [4, 5] or transition metal complexes [6, 7] are of special interest for time-resolved (TR) NEXAFS investigations. Absorption K- or L-edges of these

H. Stiel (✉) · D. Engel · J. Tümmler
Max-Born-Institute, Max-Born Strasse 2A, 12489 Berlin, Germany
e-mail: stiel@mbi-berlin.de

H. Stiel · L. Glöggler · A. Jonas · I. Mantouvalou · J. Tümmler
Berlin Laboratory for Innovative X-Ray Technologies (BLiX), Hardenbergstrasse 36, 10623 Berlin, Germany

L. Glöggler · A. Jonas · I. Mantouvalou
Institut für Optik und Atomare Physik, Technische Universität Berlin, Hardenbergstrasse 36, 10623 Berlin, Germany

M. Kozlová and J. Nejdl (eds.), *X-Ray Lasers 2018*, Springer Proceedings in Physics 241, https://doi.org/10.1007/978-3-030-35453-4_22

compounds cover the photon energy region between 100 and 1500 eV. The relevant time scales for processes in these compounds accessible by TR-NEXAFS ranging from few fs (e.g. ligand metal charge transfer [8]) up to ns (e.g. photo-dissociation [9]). Until now TR-NEXAFS investigations with high spectral and temporal resolution were mainly performed at synchrotron slicing beamline or free-electron lasers (FEL) [10]. TR-NEXAFS investigations using laboratory-based soft X-ray source were restricted either to fs time resolution at relatively low photon energies between 60 and 300 eV using high harmonics (HHG) sources [8, 11] or sub-ns time resolution in the 530 eV range using laser-produced plasma sources (LPP) [12].

In this contribution, we describe an experimental setup for TR-NEXAFS investigations covering the photon energy range from 100 eV up to 1500 eV. Using a chirped pulse amplification (CPA) pump laser with a pulse duration of 1.5 ps and an optimized spectrometer setup with a time resolution in the 10 ps region is feasible.

22.2 Experimental Setup

The experimental setup for our laboratory X-ray absorption spectrometer is shown in Fig. 22.1. It consists of a laser-based plasma source pumped by a chirped pulse amplification (CPA) thin-disk laser system, a two-channel reflection zone plate spectrometer and a back-illuminated cooled CCD detector.

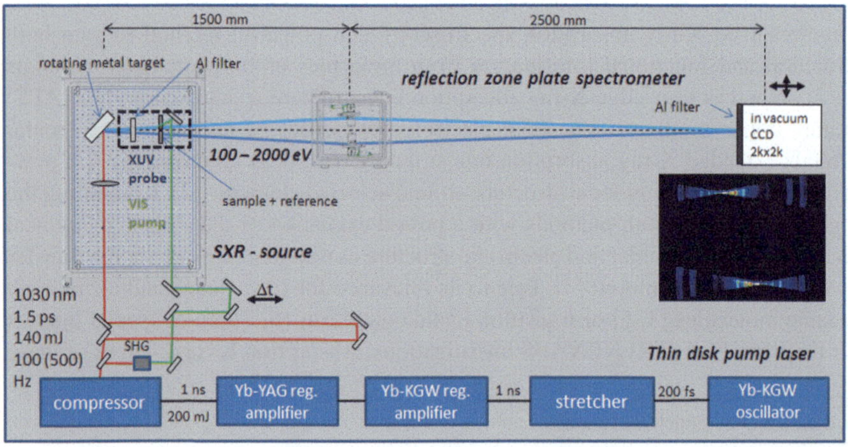

Fig. 22.1 Experimental setup

22.2.1 Laser System

As pump source, we use a chirped pulse amplification (CPA) laser system with an Yb:YAG based regenerative thin-disk amplifier as the main amplifier. It generates laser pulses with 140 mJ pulse energy at a 100 Hz repetition rate. The pulse duration can be chosen between 2 ps, 100 ps, and 1 ns.

The whole laser system consists of a Yb:KGW oscillator that is running at 80 MHz repetition rate. The oscillator is tuned to emit pulses in the spectral region around 1030 nm and bandwidth below 10 nm. The pulses are stretched in a grating stretcher to 1 ns per 1 nm bandwidth. The stretched pulses are pre-amplified in a Yb:KGW regenerative amplifier to a pulse energy of 0.5 mJ. The Yb:YAG regenerative thin-disk amplifier increases the pulse energy further to a level of more than 180 mJ with energy stability of better than 0.3% (rms) measured over several hours of operation. The amplifier medium is a 500 μm thick Yb (7%) doped YAG disk with 14 mm diameter. The thin-disk is pumped by two fibre-coupled laser diodes (Ferdinand-Braun Institute, FBH Berlin) each emitting 1 kW peak power with 10% duty cycle (1 J pulse energy in 1 ms pulses at 100 Hz repetition rate).

During the amplification process, the spectral bandwidth is narrowed to 1 nm resulting in an 1 ns uncompressed pulse duration. The grating compressor is designed for three different pulse durations <2 ps (fully compressed), 100 ps, and 1 ns (uncompressed). Including the spatial filter for cleaning the beam, the compressor has a transmission of 75%. The measured pulse duration after the compressor amounts to 1.5 ps. More details for a very similar laser system can be found in [13].

22.2.2 Laser Produced Plasma Source

The output of the thin disk laser (TDL) is divided by a beam splitter in two parts. About 95% of the beam energy is focused by a spherical lens (150 mm focusing length) on a rotating metal cylinder whereas the remaining 5% of the beam energy is converted into the second harmonic (SHG) using a BBO crystal. The SHG output with an energy of about 2 mJ is used as a pump pulse for time-resolved measurements. After passing a delay line (maximum delay 1.3 ns) the pump pulse hits the sample in a non-collinear geometry. The maximum energy flux on the sample amounts up to 10 mJ/cm^2.

The target is a rotating metal cylinder with a diameter of about 50 mm and a length of 100 mm mounted on a high precision translational stage. Both rotation speed and feed motion are fully computer controlled. We choose tungsten and copper as target materials. The laser intensity in the target plane amounts to 10^{16} W/cm^2. The X-ray emission from both target materials covers an energy range from 100 eV up to 1500 eV with an emission maximum in the 1.2 keV range. This behaviour is consistent with a model assuming that the X-ray radiation is mainly due to thermal emission from an overdense plasma with an electron temperature of about 500 eV. The X-ray

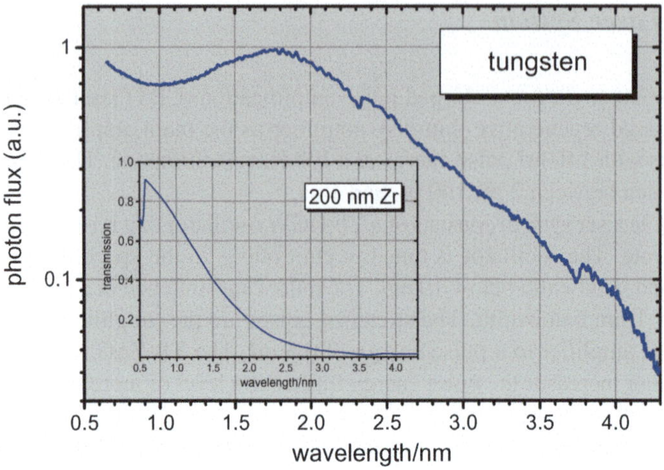

Fig. 22.2 X-ray emission from the laser-produced tungsten target. Laser: pulse duration 1.5 ps, pulse energy 100 mJ, repetition rate: 100 Hz. The spectrum was recorded using a flat field spectrometer with a HITACHI grating (part # 001—0450). Please note that the decrease of intensity in the long-wavelength region is caused by a Zr filter (thickness 200 nm) which was placed between source and spectrometer

emission spectrum of the source using a tungsten target is shown in Fig. 22.2. To protect the sample and spectrometer from debris a thin (900 nm) Mylar foil is placed at a distance of 250 mm from the source.

22.2.3 Reflection Zone Plate Spectrometer

We use off-axis reflection zone plates (RZP) as dispersive elements in our spectrometer. RZPs are two-dimensional laminar gratings with grating lines following an elliptical shape [14]. The advantage of reflection zone plates in comparison with conventional gratings is its ability to focus and disperse radiation from a point source onto the image plane of the CCD with high efficiency. RZPs are designed for a certain design wavelength. For this design wavelength, the RZP forms an image of the source in the image plane. Other wavelengths are defocused forming slightly curved lines in an X-shaped area. The energy resolution of an RZP setup is mainly determined by the source size, the geometry and the structure size of the zone plates. Choosing appropriate design parameters for a source diameter below 35 μm a spectral resolution $E/\Delta E > 1500$ is feasible [15]. In our spectrometer, we use four RZPs (Nano optics, Berlin) with design wavelengths for the O K-edge, the L-edges of Co and Ni and the MgK-edge covering the energy range from 500 eV up to 1400 eV. The RZP structures (10×80 mm^2) were etched in a Si wafer and coated with an Au layer of about 15 nm thickness. The RZPs are used under a grazing incidence angle of 2°

and a reflection angle of 3.65°. For the O K-edge and Co L-edge structures a reflectivity of 10.9% and 5.7%, respectively, has been measured at the BESSY-II optics beamline (PM1). The distance between the RZP and the LPP source is 1500 mm and the image distance amounts to 2500 mm, resulting in 1.5X magnification of the source at design wavelength and a solid angle of about 10^{-5} sr. In order to measure reference and sample spectrum simultaneously, we use two identical RZPs in our setup (cp. Fig. 22.1). Typical recording times for an SNR = 100 are of the order of 300 s. This time is mainly determined by the readout time of the detector (16-bit, cooled back-illuminated CCD camera, Roper Scientific).

A sample holder placed at a distance of 300 mm from the source houses sample and reference. Samples were produced by evaporation, sputtering or spin coating on a Si_3N_4 substrate (200 nm, clear aperture 3×3 mm^2). An empty Si_3N_4 substrate is used as a reference.

22.2.4 Data Processing

The reference and sample spectra were derived from the CCD images by summing up the pixel counts perpendicular to the dispersive axis after background correction. This simple procedure was justified by simulations due to the minute curvature of the lines of equal energy. The negative natural logarithm of the ratio of the two spectra yields the absorption of the sample and constitutes the NEXAFS spectrum. The X-axis (photon energy) has been calibrated using tabulated values for Cu lines from the NIST database.

22.3 Results

To evaluate spectral resolution, recording time and sensitivity of the setup NEXAFS spectra at L-, M- and K-edges have been recorded. Figure 22.3 shows the NEXAFS spectrum of a nickel foil (200 nm) measured at the L-edge. Both edge structures, as well as the so-called 6 eV satellite, are clearly visible. This satellite above the L_3-edge probably appears due to many-body effect which leaves two correlated 3d holes [16] and is still a matter of controversy in the literature. The spectral resolution has been estimated from the L_3-edge jump. It amounts to $E/\Delta E = 400$ @ 852.7 eV.

In order to determine the sensitivity of the setup a sample containing Ta, Gd, Tb and Dy in a sandwich-like structure. The sample area was less than 1 mm^2, the thickness was 157 nm. As can be seen from Fig. 22.4 the M_4- and M_5-edges of all three lanthanides are clearly distinguishable. Further, the absolute values of the absorption coefficient derived from the measurement are in agreement with tabulated values from the HENKE database [17].

The performance of the setup for tender X-ray investigations of photosynthetic pigments containing magnesium as a central metal atom [5] has been tested using

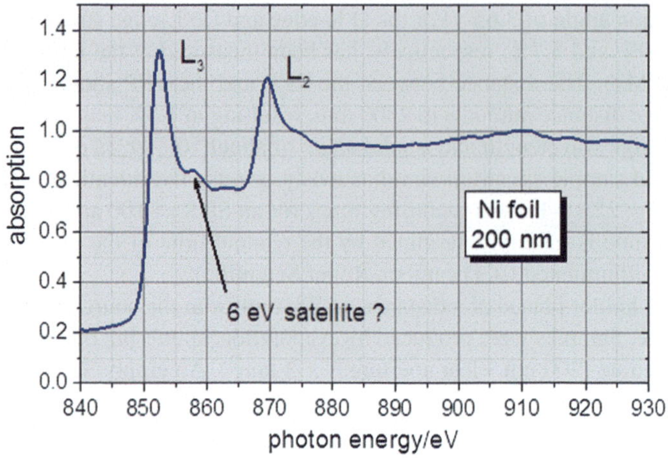

Fig. 22.3 L-edge NEXAFS spectrum of a nickel foil (thickness 200 nm) obtained with the setup depicted in Fig. 22.1

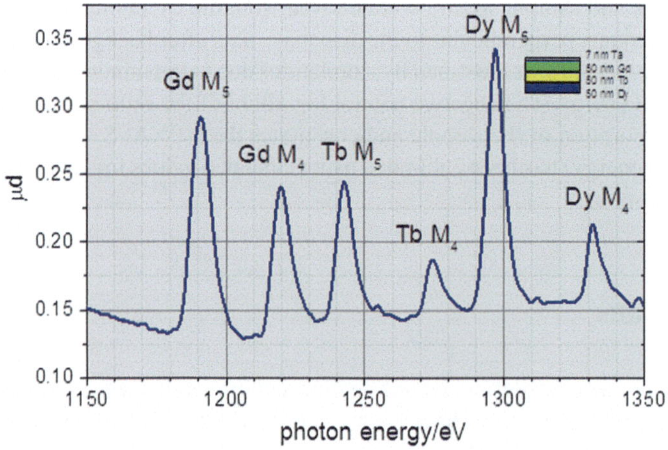

Fig. 22.4 M-edge NEXAFS spectrum of a sandwich structure consisting of 7 nm Ta, 50 nm Gd, 50 nm Tb and 50 nm Dy obtained with the setup depicted in Fig. 22.1

an Mg foil. The result is shown in Fig. 22.5. As can be seen from Fig. 22.5 besides the near edge structure at 1308 eV oscillations in the region above 1340 eV occur. These oscillations are related to an extended absorption fine structure (EXAFS).

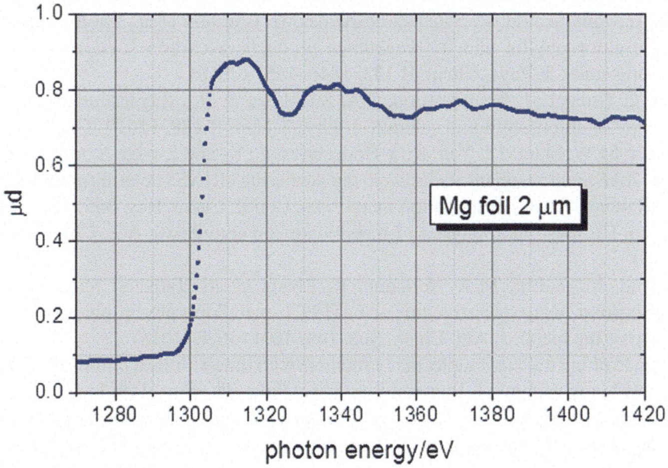

Fig. 22.5 K-edge NEXAFS spectrum of a magnesium foil (thickness 2 μm) obtained with the setup depicted in Fig. 22.1

22.4 Conclusions and Outlook

We have described a setup for laboratory-based near-edge X-ray absorption fine structure spectroscopy covering a photon energy range from 100 eV up to 1500 eV. The setup is capable to perform time-resolved X-ray absorption with a spectral resolution of some eV and a temporal resolution below 50 ps. We demonstrated the performance of the setup using different kinds of the sample with relevance for material and life sciences.

The next step is to apply the setup for time-resolved NEXAFS investigations on biomolecules in solution. To do this we have developed a wet chamber that is currently under commissioning.

Acknowledgements HS is grateful for support by LASERLAB Europe Grant Agreement #654148.

References

1. M. Chergui, Time-resolved X-ray spectroscopies of chemical systems: New perspectives. Str. Dyn. **3**, 031001 (2016)
2. A. Koide, T. Fujikawa, N. Ichikuni, Recent progress in EXAFS/NEXAFS spectroscopy. J. Electron Spectrosc. Relat. Phenom. **195**, 375–381 (2014)
3. C.J. Milne, R.M. Van der Veen, P. Van-Thai, F.A. Lima et al., Ultrafast X-ray absorption studies of the structural dynamics of molecular and biological systems in solution. Chimia **65**, 303–307 (2011)

4. K. Witte, I. Mantouvalou, R. Sánchez-de-Armas, H. Lokstein et al., On the electronic structure of Cu chlorophyllin and its breakdown products: a carbon K-edge X-ray absorption spectroscopy study. J. Phys. Chem. B **122**, 1846–1851 (2018)
5. K. Witte, C. Streeck, I. Mantouvalou, S.A. Suchkova et al., Magnesium K-edge NEXAFS spectroscopy of chlorophyll a in solution. J. Phys. Chem. B **120**, 11619–11627 (2016)
6. M.L. Baker, M.W. Mara, J.J. Yan, K.O. Hodgson et al., K- and L-edge X-ray absorption spectroscopy (XAS) and resonant inelastic X-ray scattering (RIXS) determination of differential orbital covalency (DOC) of transition metal sites. Coord. Chem. Rev. **345**, 182–208 (2017)
7. M. Chergui, Ultrafast photophysics of transition metal complexes. Acc. Chem. Res. **48**, 801–808 (2015)
8. E.S. Ryland, M.-F. Lin, M.A. Verkamp, K. Zhang et al., Tabletop femtosecond M-edge X-ray absorption near-edge structure of FeTPPCl: metalloporphyrin photophysics from the perspective of the metal. J. Am. Chem. Soc. **140**, 4691–4696 (2018)
9. L.X. Chen, Probing transient molecular structures with time-resolved pump/probe XAFS using synchrotron X-ray sources. J. Electron Spectrosc. Relat. Phenom. **119**, 161–174 (2001)
10. T.J.A. Wolf, R.H. Myhre, J.P. Cryan, S. Coriani et al., Probing ultrafast $\pi\pi^*/n\pi^*$ internal conversion in organic chromophores via K-edge resonant absorption. Nat. Commun. **8**, 29 (2017)
11. Y. Pertot, C. Schmidt, M. Matthews, A. Chauvet et al., Time-resolved X-ray absorption spectroscopy with a water window high-harmonic source. Science **355**, 264–267 (2017)
12. P. Grossmann, I. Rajkovic, R. More, J. Norpoth et al., Time-resolved near-edge X-ray absorption fine structure spectroscopy on photo-induced phase transitions using a tabletop soft-X-ray spectrometer. Rev. Sci. Instrum. **83**, 053110–053114 (2012)
13. R. Jung, J. Tummler, I. Will, Regenerative thin-disk amplifier for 300 mJ pulse energy. Opt. Express **24**, 883–887 (2016)
14. T. Wilhein, D. Hambach, B. Niemann, M. Berglund et al., Off-axis reflection zone plate for quantitative soft X-ray source characterization. Appl. Phys. Lett. **71**, 190–192 (1997)
15. I. Mantouvalou, K. Witte, W. Martyanov, A. Jonas et al., Single shot near edge X-ray absorption fine structure spectroscopy in the laboratory. Appl. Phys. Lett. **108**, 201106 (2016)
16. J. Sanchez-Barriga, J. Braun, J. Minar, I. Di Marco et al., Effects of spin-dependent quasiparticle renormalization in Fe, Co, and Ni photoemission spectra: an experimental and theoretical study. Physical Review B **85** (2012)
17. B.L. Henke, E.M. Gullikson, J.C. Davis, X-ray interactions: photoabsorption, scattering, transmission, and reflection at E = 50–30,000 eV, Z = 1–92. At. Data Nucl. Data Tables **54**, 181–342 (1993)

Chapter 23
Ptychography and Single-Shot Nanoscale Imaging with Plasma-Based Laser Sources

F. Tuitje, M. Zürch, T. Helk, J. Gautier, F. Tissandier, J.-P. Goddet, E. Oliva, A. Guggenmos, U. Kleineberg, H. Stiel, S. Sebban and C. Spielmann

Abstract We report the direct wavefront characterization of an intense ultrafast high-harmonic seeded soft X-ray laser at 32.8 nm wavelength and monitor the exit of the laser plasma amplifier depending on the arrival time of the seed pulses with respect to pump pulses. For the wavefront measurement in phase and intensity, we used high-resolution ptychography. After propagating the wavefront back to the source, we are able to observe the rear end of the plasma amplifier. We compare the characteristics of the seeded soft X-ray Laser to an unseeded one and find an increasing beam stability and lateral coherence important for lensless imaging techniques.

F. Tuitje · M. Zürch (✉) · T. Helk · C. Spielmann
Institute for Optics and Quantum Electronics, Abbe Center of Photonics, University of Jena, Jena, Germany
e-mail: michael.zuerch@uni-jena.de

Helmholtz Institute Jena, Jena, Germany

M. Zürch · A. Guggenmos
Department of Chemistry, University of California at Berkeley, Berkeley, USA

M. Zürch
Fritz Haber Institute of the Max Planck Society, Berlin, Germany

J. Gautier · F. Tissandier · J.-P. Goddet · S. Sebban
Laboratoire d'optique appliquée, ENSTA-ParisTech, Palaiseau, France

E. Oliva
Departamento de Ingeníeria Energética, ETSI Industriales, Universidad Politécnica de Madrid, Instituto de Fusión Nuclear, Universidad Politecnica de Madrid, Madrid, Spain

A. Guggenmos · U. Kleineberg
Department for Physics, Ludwig-Maximilian-University Munich, Garching, Germany

H. Stiel
Max-Born Institute Berlin, Berlin, Germany

M. Kozlová and J. Nejdl (eds.), *X-Ray Lasers 2018*, Springer Proceedings in Physics 241, https://doi.org/10.1007/978-3-030-35453-4_23

23.1 Introduction

In the extreme ultraviolet (XUV) to the soft X-ray regime emitting plasma-based sources relying on amplified spontaneous emission gain popularity due to their table-top form factor and increasing flux which makes them comparable to large facilities like synchrotrons. Due to their high photon count in a narrow bandwidth, these soft X-ray lasers (SXRL) match the requirements for high-resolution imaging applications [1, 2]. Especially lensless imaging applications, such as coherent diffraction imaging (CDI) or ptychography, rely on the coherence of the illuminating source, which is limited for SXRL.

On the other hand, XUV sources based on high harmonic generation (HHG) have excellent spatial and temporal coherence and are therefore widely used for imaging [3, 4]. However, the rather low flux of HHG sources limits the resolution. To overcome the limitations in coherence and flux, seeding an SXRL with HHG radiation will result in an extraordinary high flux and excellent coherence with a compact setup [5]. Questioning the spatial coherence and the flux of individual XUV pulses emitted from SXRLs, we performed single shot CDI measurements with an HHG-seeded SXRL at the Laboratoire d'optique appliquée (LOA) and compare them to previous measurements with an unseeded SXRL at Max-Born Institute. Additionally, we derived the experimental parameters for ptychographic imaging of known samples for retrieving the complex wavefront at the focus position. By propagating the measured illumination function back to its source, we can monitor the complex exit field of the seeded plasma amplifier. By varying the delay between seed and pump beam, plasma dynamics become observable by the usage of state-of-the-art imaging techniques.

23.2 Experimental Setup

The experiment was performed at LOA using the 2 J "Salle jaune" Ti:Saphire laser system [5]. As seen in Fig. 23.1 the main pulse is split into sub pulses: an ignitor pulse followed by a heater pulse with a duration of 600 ps is focused with an axicon into a Krypton gas jet forming a wave guiding plasma channel. Pumped by a 1.36 J, 30 fs long pump pulse, this plasma channel acts as the laser gain medium and obtain population inversion through optical field ionization of the nickel-like Krypton (Kr^{8+}) ions [6]. Beyond the scheme of amplified spontaneous emission, the SXRL was seeded by an HHG pulse at 32.8 nm, matching the laser transition in Kr^{8+}, obtained by focusing a 16 mJ, 350 fs, 813 nm pulse into argon. The time delay between seed pulse and pump pulse is variable through a delay line. Seeding of the SXRL allows creating short pulses in the extreme ultraviolet with around 450 fs pulse duration at 32.8 nm with a relative bandwidth $\lambda/\Delta\lambda$ of ~10,000 and up to 3×10^{11} photons per shot with a repetition rate up to 10 Hz [5, 7].

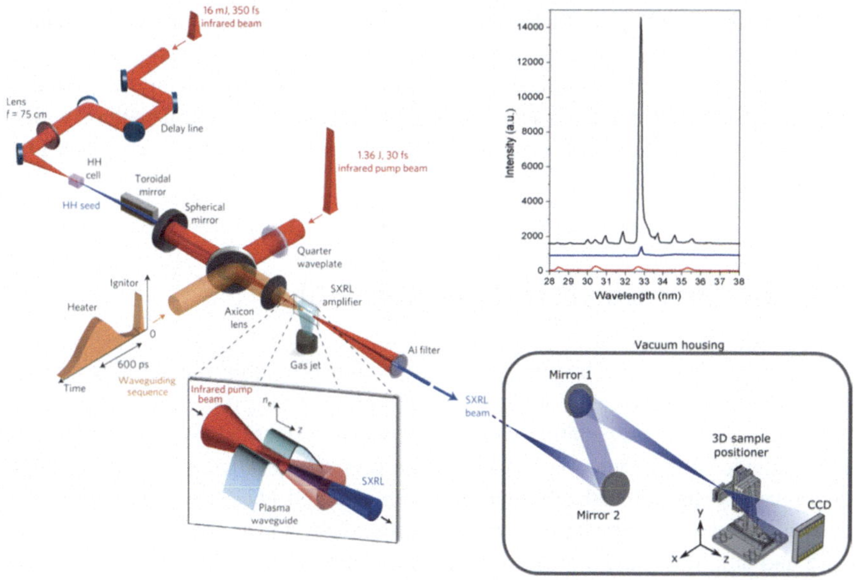

Fig. 23.1 Experimental setup consisting of the SXRL [5] (left) and the ptychography/CDI system. The temporal delay between the pump and seed pulses is adjusted by an optical delay line. After filtering of the XUV radiation with aluminum filters, the beam is focused on the sample mounted on a high-precision 3D positioner. The resulting diffraction patterns are captured with a CCD camera. The graph shows the spectra for HHG (red), typical ASE (blue) and the seeded SXRL (black)

After exiting the plasma and blocking the residual infrared radiation with aluminum filters, the XUV beam propagates 5 m to the first multilayer mirror (mirror 2). Then, the beam is focused onto the sample by two spherical multilayer mirrors with 5 m (mirror 2) and 0.5 m (mirror 1) focal length, respectively, resulting in a demagnification of 10. To reduce aberration, the angle of incidence on the curved mirrors was minimized to 4°.

To implement ptychographic scanning, the sample is mounted on a 3D positioner with a repeatability of 50 nm over the complete travel range of 30 mm. The intensity distribution of diffraction patterns was recorded with a 2048 × 2048 pixel CCD detector with 16-bit dynamic range and 27.8 mm diagonal size (Andor iKon-L SO). The camera was cooled down to −50 °C to reduce noise. With the distance between sample and camera of 55 mm, we realized a numerical aperture (NA) of 0.25.

For the reconstruction of the sample from the single-shot diffraction patterns, we used the hybrid-input–output algorithm. Note for the code we set the feedback parameter β to 0.9 and used a shrink wrap support [8]. The ptychographic reconstruction was obtained with an implementation of the extended ptychographic iterative engine (ePIE) [9]. It is capable of reconstructing the object with an automatically adapting field of view and the illumination function (also called probe) in phase and intensity. It is worth to mention that the reconstructed probe is the mean of all probes at every scanning point. The feedback parameter α and β for object and probe were chosen to

0.9. Due to the coupling of the feedback functions, the probe reconstruction cannot directly start with the object reconstruction. A delay of 10 iterations avoids artifact oscillations between probe and object fields. As a further parameter, we assumed a 6 μm flat-top probe and 30 scanning points arranged in a spiral pattern with a 90% overlap to avoid artifacts caused by the computing grid, known as "raster grid pathology" [10].

23.3 Backpropagation of Recovered Fields

Due to the complete ptychographic reconstruction providing the complex-valued illumination function on the sample position, knowing the properties of the optical components and the propagation length in between, numerical propagation back to the exit plane of the source is possible. As a propagation method, we use a Fresnel propagator based on the angular spectrum method [11]. According to higher bandwidth, we oversampled the reconstructed fields by a factor of 4 by interpolating phase and amplitude and increased the field of view (FOV) to 220 μm by padding the complex matrices with zeros. Therefore, we can ensure a lateral extension of the field beyond the FOV determined by the reconstruction.

The first propagation of 0.5 m takes place from the measured focus to the first mirror M1 with a focal length of 0.5 m, see Fig. 23.2.

The mirror curvature is considered by adding a phase shift. The same procedure is repeated at the second mirror M2 with a focal length of 5 m after propagation of 0.3 m between the two mirrors. The results after the final propagation from M2 to the

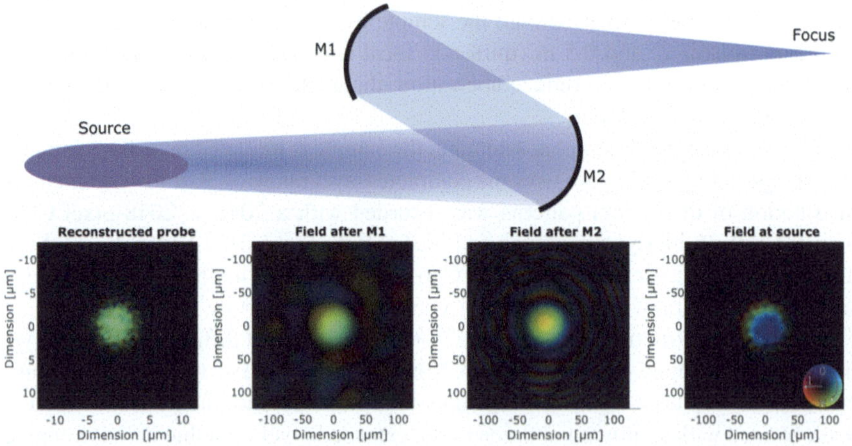

Fig. 23.2 Schematic experimental setup with corresponding complex fields. The reconstructed probe is at a focus for 1.2 ps delay (left) after propagation to the first mirror M1. After further propagation over M2 to the source (right), the structure of the exit wave reveals. Note the expanding FOV during the propagation. Phase and Intensity are colour/brightness coded (see inset)

source exit reveal the complex structure of the exit wave. Note, this method allows determining the shape of the wavefront with high precision at the exit of the laser plasma amplifier. Furthermore, the backpropagation only considers monochromatic radiation, the image of the source's exit fields are therefore the only representative for the wavelength used for the ptychographic reconstruction.

23.4 Experimental Results and Image Reconstruction

We captured single-shot diffraction patterns of different samples with ~450 fs pulse duration. We use the full chip size for the images and obtained a peak intensity of 7500 counts after subtracting the background.

As seen in the lineout in Fig. 23.3. a resolution between 140 and 250 nm was achieved. Compared to the theoretical resolution limit of ~70 nm, the lower resolution is explained by the low signal strength at higher spatial frequencies. A series of single-shot images revealed high repeatability of reconstruction and, therefore, low variations in subsequently following single-shot measurements. This is in strong contrast to the results of similar imaging experiments with a non-seeded SXRL [1]. Assuming the emitting diameter of the plasma channel to be ~80 to ~100 μm [5], we can estimate the full illuminated area on the sample to 8–10 μm (FWHM). The coherent fraction has been estimated to 5–6 μm. This was obtained with CDI reconstructions of a well-characterized hole pattern (Quantifoil Micro Tools, R1/2) with hole diameters of 1 μm and hole-to-hole distances of 2 μm.

Therefore, the coherent area represents up to 40% of the overall illuminated area. This is in contrast to results with the non-seeded SXRL [1], where the ratio of coherent diameter to the overall beam diameter is 6–9 times lower.

For the ptychographic measurement, we used again the above-described hole pattern. With the calculated size of the coherent illumination spot from the previous

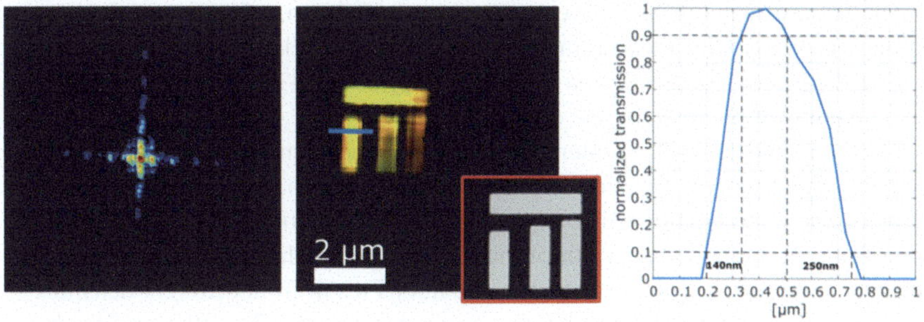

Fig. 23.3 Single-shot diffraction pattern (left), reconstructed object (center) and lineout profile (right) at the position of the blue bar. Using the 10/90 criterion, a resolution between 140 and 250 nm was achieved. The colour-coding of the objects corresponds to Fig. 23.2. The inset shows an SEM image of the binary sample consisting of an ion beam structured gold-coated Si_3N_4 membrane of 200 nm thickness

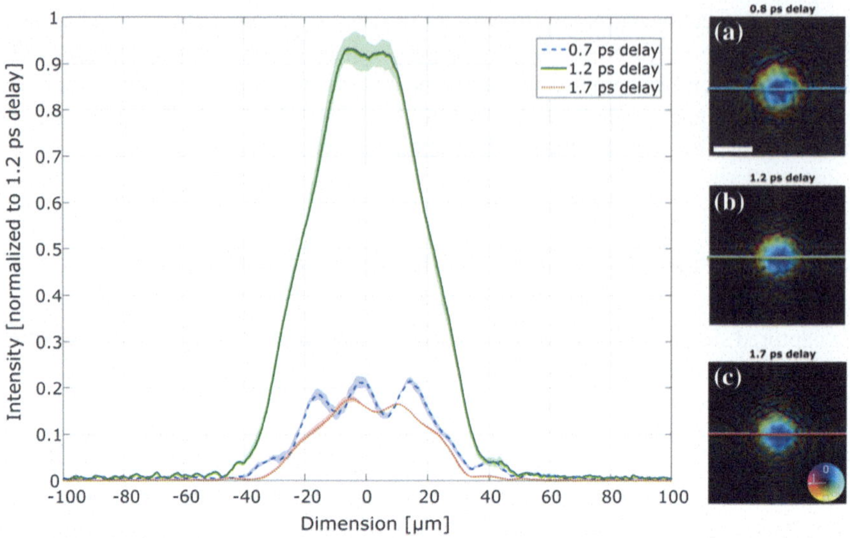

Fig. 23.4 Intensity lineouts of the reconstructed and backpropagated probes for different delays of 0.8 ps (**a**), 1.2 ps (**b**) and 1.7 ps (**c**), respectively. The lineouts are normalized to the maximum signal for 1.2 ps delay. The corresponding errors are marked with pale colours. The low-frequency modulations can be explained by saturation effects in the plasma. The white scale bar in (**a**) represents 50 μm. The colour-coding as seen in (**c**): colour represents phase and brightness represents intensity

measurement, designing a map with sufficient overlap of 90% was possible. We took single-shot ptychographic datasets for seed-pump-delays of 0.8, 1.2, and 1.7 ps under the same conditions as in CDI. A delay of 1.2 ps was found as the optimum in terms of total output yield. To reduce noise and statistical variations in the beam profile, we averaged five measurements per delay, ensuring a stable wavefront image. Together with averaging over 30 scanning points, every reconstructed wavefront per delay corresponds to an average of 150 single-shot patterns (Fig. 23.4).

The reconstructed probes have a diameter of ~6.5 μm, which matches the first estimation of the CDI measurement. After applying backpropagation, the optical fields at the position of the plasma channel exit increased in size to ~70 μm, matching the demagnification given by the optics.

The compared normalized intensities reflect the optimum delay of 1.2 ps regarding the total output power. Lineouts of the intensities reveal Gaussian-like profiles with low-frequency modulations. The high-frequency modulations are an artifact stemming from the limited bandwidth of the propagation method and have no physical meaning, whereas the low-frequency modulations result from different densities of the lasing nickel-like Kr^{8+} ion. For the optimal delay of 1.2 ps, the profile shows a dip at the center, explainable by over ionization at the waveguide center. The 0.8 and 1.7 ps profiles show an overall lower intensity due to non-optimal timing for the best amplification. Furthermore, the profiles show more complex modulations of the exit wave, pointing out changes in the lasing ion abundance. In this case, the seed beam corresponds to a probe beam in a classical pump-probe experiment. The seed interrogates the lasing ion density in the plasma at the time of beam arrival. Assuming a relaxation time of the highly ionized gas of several hundred picoseconds to nanoseconds, the retrieved image of the exit wave can be seen as a non-averaging snapshot of the plasma density.

23.5 Conclusion

Besides testing the characteristics of an HHG-seeded SXRL concerning coherence and flux using single-shot diffraction pattern, a method was shown to monitoring the exit of the plasma channel by back-propagating reconstructed focus illuminations back to its source. By tuning the delay between pump and seed pulse, the saturation dynamics of the plasma channel become visible. A comparison to an unseeded SXRL based on solid-state molybdenum target [1] shows significant higher wavefront stability and an improved spatial coherence up to 75%. The high flux around 3×10^{11} photons per shot allows single-shot imaging without the necessity of accumulating shots under incoherent superposition. Pulse duration down to 450 fs enables ultrafast imaging of dynamic processes at the highest resolutions. Due to the low shot-to-shot variations, ptychography becomes easily realizable with scanning parameters extracted from the CDI measurements. Due to the ability of the ePIE algorithm to reconstruct object and probe simultaneously, phase and intensity of the focus could be characterized. Assuming known optics, a propagation of the reconstructed probe back to its origin at the end of the plasma amplifier reveals the structure of the exit wave. Varying the delay between pump and seed from 0.8 to 1.7 ps, saturation dynamics of the lasing ion were imaged.

Acknowledgements The research leading to these results has received funding from LASERLAB-EUROPE (grant agreement no. 654148, European Union's Horizon 2020 research and innovation programme) and through a research group in the State of Thuringia funded by the European social fund (2015 FGR 0094). Furthermore, the authors acknowledge the QUEST for ENERGY project funded by the Federal Ministry of Education and Research (BMBF) under "Make our Planet Great Again—German Research Initiative", Grant No. 57427209, implemented by the German Academic Exchange Service (DAAD). M. Z. acknowledges the Alexander von Humboldt association.

References

1. M. Zürch et al., Transverse coherence limited coherent diffraction imaging using a molybdenum soft X-ray laser pumped at moderate pump energies. Sci. Rep. **7**(1), 5314 (2017)
2. H.C. Kang et al., Single-pulse coherent diffraction imaging using soft X-ray laser. Opt. Lett. **37**(10), 1688–1690 (2012)
3. M.D. Seaberg et al., Ultrahigh 22 nm resolution coherent diffractive imaging using a desktop 13 nm high harmonic source. Opt. Express **19**(23), 22470–22479 (2011)

4. M. Zürch, C. Kern, C. Spielmann, XUV coherent diffraction imaging in reflection geometry with low numerical aperture. Opt. Express, OE **21**(18), 21131–21147 (2013)
5. A. Depresseux et al., Table-top femtosecond soft X-ray laser by collisional ionization gating. Nat. Photonics **9**(12), 817–821 (2015)
6. S. Sebban et al., Demonstration of a Ni-like Kr optical-field-ionization collisional soft X-ray laser at 32.8 nm. Phys. Rev. Lett. **89**(25), 253901 (2002)
7. P. Zeitoun et al., A high-intensity highly coherent soft X-ray femtosecond laser seeded by a high harmonic beam. Nature **431**(7007), 426–429 (2004)
8. S. Marchesini et al., X-ray image reconstruction from a diffraction pattern alone. Phys. Rev. B **68**(14), (2003)
9. A.M. Maiden, J.M. Rodenburg, An improved ptychographical phase retrieval algorithm for diffractive imaging. Ultramicroscopy **109**(10), 1256–1262 (2009)
10. P. Thibault, M. Dierolf, O. Bunk, A. Menzel, F. Pfeiffer, Probe retrieval in ptychographic coherent diffractive imaging. Ultramicroscopy **109**(4), 338–343 (2009)
11. T. Kozacki, K. Falaggis, Angular spectrum-based wave-propagation method with compact space bandwidth for large propagation distances. Opt. Lett. **40**(14), 3420–3423 (2015)

Chapter 24
Pico–Femto–Attosecond Metrology for Advanced and XUV Photon Sources

T. Sinyakova, C. Bomme, D. Cubaynes, C. Bourassin-Bouchet, A. Gharbi, G. Garcia, A. K. Pandey, I. Papagiannouli, O. Guilbaud and A. Klisnick

Abstract We present a new Velocity Map Imaging (VMI) spectrometer based on the thick-lens configuration to be implemented at different XUV source beamlines to realize the measurements of the pulse temporal structure using the technique of laser-dressed photoionization.

24.1 Introduction

The temporal characterization of ultrashort XUV pulses generated from seeded plasma-based XUV lasers [1], high-order harmonics from plasma mirrors [2], or free-electron lasers [3] is a crucial challenge for both the ongoing development of these advanced sources and for the interpretation of their interaction with matter in various applications, such as ultrafast dynamics in atoms and molecules, or multiphoton processes [4, 5]. There is thus an important worldwide effort toward establishing adapted, reliable, and easy-to-use techniques of temporal metrology in the femtosecond to attosecond range, possibly on a single-shot basis, using different approaches [6–8].

T. Sinyakova · D. Cubaynes · A. Gharbi · A. Klisnick (✉)
CNRS, ISMO, Paris-Sud University, Orsay, France
e-mail: annie.klisnick@u-psud.fr

C. Bomme
CEA-Saclay, LIDYL, Saclay, France

C. Bourassin-Bouchet
Institut of Optique, LCF, Palaiseau, France

G. Garcia
Synchrotron SOLEIL, Saint-Aubin, France

A. K. Pandey · I. Papagiannouli · O. Guilbaud
LPGP, CNRS, Paris-Sud University, Orsay, France

M. Kozlová and J. Nejdl (eds.), *X-Ray Lasers 2018*, Springer Proceedings in Physics 241, https://doi.org/10.1007/978-3-030-35453-4_24

In the frame of a collaborative project led by ISMO and LCF and gathering several XUV source developers within Paris-Saclay University, we are currently developing a new portable instrument, based on a home-made Velocity Map Imaging (VMI) spectrometer. This device was designed following the thick-lens configuration proposed by Kling et al. [9] to allow higher resolution of the photoelectrons over a broader range of kinetic energy compared to standard VMI designs. This VMI spectrometer will be implemented at different laser-based XUV source beamlines to conduct measurements of the pulse temporal structure using the technique of laser-dressed photoionization, already demonstrated elsewhere. Depending on the relative timescale between the duration of the ionizing XUV pulse and the period of the dressing laser field, both the sidebands [10] and the streaking regime [11] will be investigated. The main advantage of using a VMI spectrometer rather than other types of photoelectron spectrometers [(time-of-flight (TOF), magnetic bottle (MB)] is the higher collection efficiency and the access to angularly resolved distributions of photoelectron velocities, from which temporal information could be extracted.

In this paper, we discuss the first step of our ongoing project devoted to the construction and commissioning of our new VMI spectrometer. We have used two types of sources, (i) a conventional helium lamp, (ii) an ASE plasma-based XUV laser operated at 32.6 nm available at the LASERIX facility. Tests with different rare gases were performed to characterize the performance of the spectrometer in terms of sensitivity and resolution. The measurements were compared with numerical simulations using the SIMION code.

24.2 VMI Spectrometer

The principle of the VMI spectrometer consists in projecting the expanding sphere of photoelectrons produced by the interaction of a gas jet with a XUV beam onto a 2D detector with an intense electrostatic field applied across the electrodes as shown in Fig. 24.1. The photoelectrons with the same velocity vector are projected onto the same position on the detector, independently of their point of emission in the interaction region. Focusing effect is produced by the thick-lens configuration involving 11 electrodes including a repeller, extractor, and focusing electrodes [9]. This configuration allows optimization of the spectrometer resolution over a broad kinetic energy range $E \sim 0$–100 eV at the detector plane with full angular information. The gas is injected by the polarized needle going through the ⌀ 3 mm hole in the repeller. Then, the photoelectrons are detected by a single multichannel plate coupled to a phosphor screen to convert the electron signal into luminescence which is observed by a CCD camera.

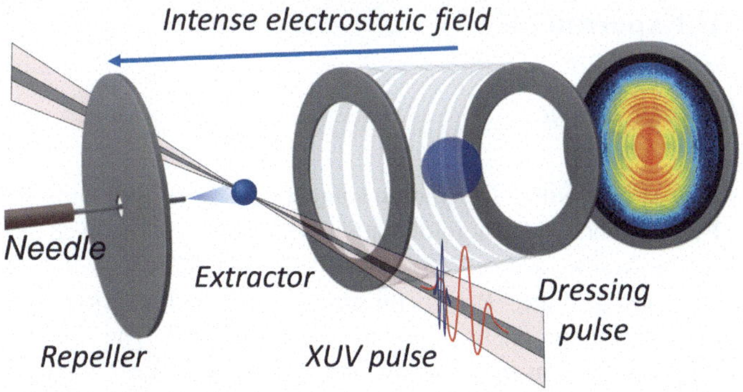

Fig. 24.1 Scheme of VMI spectrometer with thick-lens configuration

24.3 Numerical Simulations

Numerical simulations of the trajectories of photoelectrons in the electrostatic field were performed using SIMION code to determine the expected resolution and range of electron's kinetic energy as well as to optimize the voltages of the electrodes. The photoelectron trajectories were calculated for different relevant kinetic energies and voltages of repeller. Their positions and radius R on the detector plane are presented in Fig. 24.2. The resolution of the spectrometer is estimated between 2 and 5%, depending on the applied voltages and energy range.

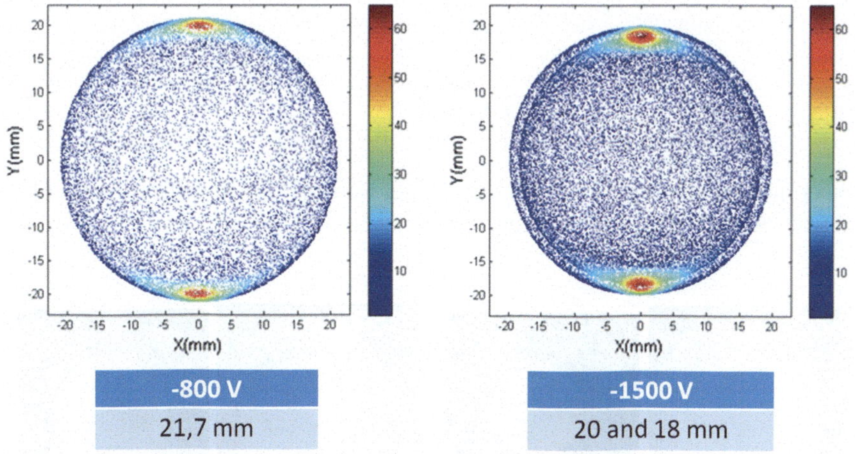

Fig. 24.2 Simulated images of photoelectrons for $E_k = 5.45$ eV (left side) and $E_k = 9.08$ and 7.77 eV (right side) and the radius R

24.4 VMI Experiments

24.4.1 Helium Lamp

First tests were realized with a DC photon source, namely helium lamp with energy $E_{h\upsilon} = 21.21$ eV. Valence np^6-photoionization produces an unresolved electron line of 5.45 eV in the Argon case ($n = 3$) and of 9.08 and 7.77eV in the Xenon case ($n = 5$) corresponding to the $^2P_{3/2,1/2}$ thresholds. Raw and inverted photoelectron images for relevant kinetic energies are shown in Fig. 24.3. The doublet states of Xe^+ can be distinguished in the both raw and inverted images. Comparison of the measured R with simulated results for both gases and different voltage settings shows good agreement.

The photoelectron spectra obtained from the inverted images allow to determine the resolution of the VMI spectrometer. Each peak of these spectra was fitted with a Gaussian profile. The energy resolutions $\Delta E/E$ are 8.3 and 10% in the Ar case and 8.4 and 9.6% in the Xe case for the mentioned voltages of the repeller. The experimentally found energy resolution is lower than was predicted by the simulation due to a larger size of the interaction zone (1.5 × 4 mm) given by an intersection between the gas jet and the diaphragmed diverging photon beam since the light was not focused in the VMI, but more than satisfactory for future tests.

The calibration curve of the VMI spectrometer is given by following expression: $E_k = V_{rep}\frac{R^2}{C^2}$, where C is the VMI constant and R is the measured radius, and illustrated in Fig. 24.4 [13]. A linear fit to the measured results provides a good representation of the $E(R^2)$ dependence and the coefficient of the linear fit k gives the value of the VMI constant C: $k = 1/C^2$. The constant C is equal to 25.9 for the experimental data and 25 for the SIMION simulations for the current geometry.

24.4.2 XUV Plasma-Based Laser

Experiments with the plasma-based XUV laser were performed with Ne target. Photons of energy of 38 eV produce an unresolved electron line of $E_k = 16.4$ eV asso-

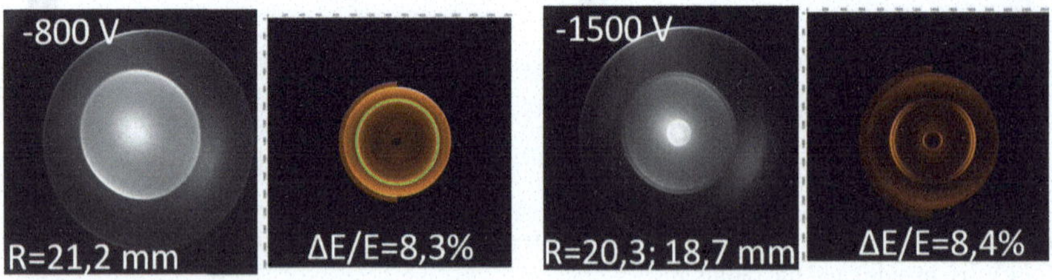

Fig. 24.3 Raw and inverted images of photoelectrons of $E_k = 5.45$ eV (left side) and of $E_k = 9.08$ and 7.77 eV (right side)

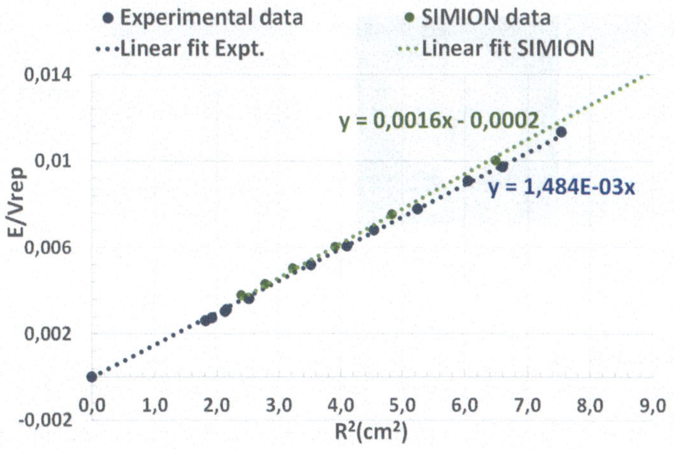

Fig. 24.4 Experimental (blue) and simulated (green) energy calibration curves $E_k/V_{rep} = f(R^2)$ as well as its linear fit

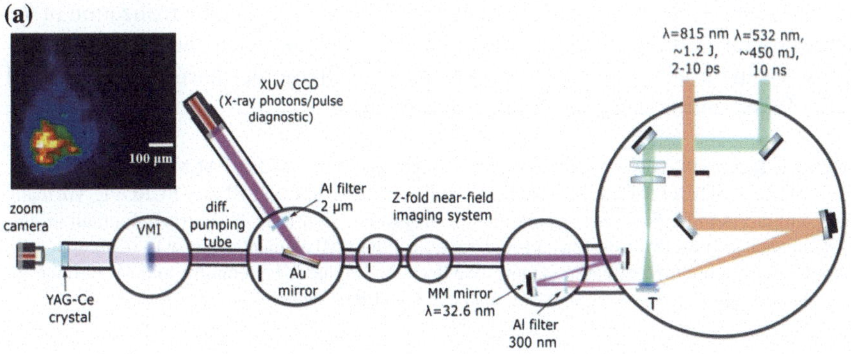

Fig. 24.5 Schematic of the setup of the plasma-based X-ray laser. Inset figure **a** depicts XUV laser focused at the VMI center

ciated with the $^2P_{3/2,1/2}$ thresholds. The experimental scheme of the solid target plasma-based soft X-ray LASERIX source is shown in Fig. 24.5 [12]. The exit of Ne-like Ti X-ray laser plasma is imaged with a magnification of ~10 into VMI spectrometer. For diagnostics of the X-ray source size, position, and energetic characteristics, an X-ray CCD was placed approximately at the same path-length as for the center of the spectrometer. The X-ray spot size into the VMI spectrometer is 190 × 260 μm with a standard deviation of 11% between two X-ray laser pulses.

The left and right halves of Fig. 24.6 show the measured raw and inverted images. The energy resolution $\Delta E/E$ found from the photoelectron spectra for different voltages is 8–10%. Despite the smaller size of the interaction zone, the found resolution is lower than predicted by the SIMION simulations and need further optimizations. The single-shot capability of VMI spectrometer was also successfully tested. The raw VMI image taken in single-shot at 38 eV is shown in Fig. 24.6.

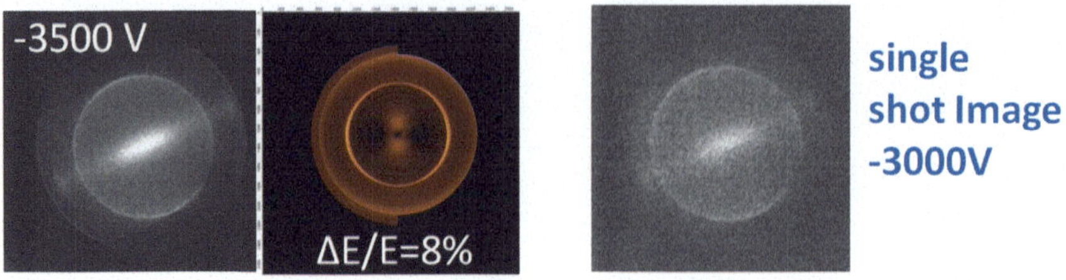

Fig. 24.6 Raw and inverted images of photoelectrons of $E_k = 16.4$ eV

24.5 Conclusion

Our new VMI spectrometer has been implemented and characterized at LASERIX (LUMAT, Université Paris-Sud, France) for commissioning with a helium lamp and a plasma-based XUV laser pulse. The SIMION simulations were performed to confirm the reliability of the spectrometer. The found energy resolution is actually lower than was predicted by the numerical simulations but satisfactory for the realization of the future program. Further, the spectrometer will be implemented at the UHI100 facility (CEA-Saclay, France) for the characterization of high-order harmonics generated from plasma mirror, using an IR laser dressing field.

Acknowledgements This project has received funding from LabEx PALM (ANR-10-LABX-0039-PALM), LIDEX OPT2X, Sesame Ile de France PULSE-X, LASERLAB-EUROPE. We thank ISMO engineers and technicians, in particular, J. Guigand, C. Charrière, N. Tournier, for their invaluable technical support in the design and construction of the VMI elements. One of us (CB) gratefully acknowledges support from Fabien Quéré. We thank E. Baynard, J. Demailly, and M. Pittman for their support during the experiments conducted at LASERIX facility.

References

1. A. Depresseux et al., Nat. Photonics **9**, 817 (2015)
2. C. Thaury, F. Quéré, J. Phys. B At. Mol. Opt. Phys. **43**, 213001 (2010)
3. E. Allaria et al., Nat. Photonics **6**, 699 (2012)
4. M. Richter, S.V. Bobashev, A.A. Sorokin, K. Tiedtke, J. Phys. B At. Mol. Opt. Phys. **43**, 194005 (2012)
5. F. Bencivenga, F. Capotondi, E. Principi, M. Kiskinova, C. Masciovecchio, Adv. Phys. **63**, 327 (2015)
6. S. Duesterer et al., Phys. Rev. Spec. Top-Ac. **17**, 120702 (2014)
7. A. Azima et al., New J. Phys. **20**, 013010 (2018)
8. Y. Nomura et al., Nat. Photonics **5**, 124 (2009)
9. N.G. Kling et al., J. Instrum. **9**, P05005
10. S. Haessler et al., J. Phys. B At. Mol. Opt. Phys. **45**, 074012 (2012)
11. U. Frühling et al., Nat. Photonics **3**, 523 (2009)
12. O. Delmas et al., Opt. Lett. **39**, 6102 (2014)
13. G.A. Garcia et al., Rev. Sci. Instrum. **76**, 053302 (2005)

Chapter 25
High Spectral and Spatial Resolution Soft X-ray/XUV VLS Spectrographs

E. N. Ragozin, A. O. Kolesnikov, A. S. Pirozhkov, P. V. Sasorov, A. N. Shatokhin and E. A. Vishnyakov

Abstract We report the implementation of three high-resolution VLS-grating spectrometers intended for laboratory applications: (i) an inherently stigmatic (imaging) broadband (12.5–30 nm) spectrograph, which uses the synergy of an aperiodic normal-incidence multilayer mirror and a plane grazing-incidence VLS grating; (ii) a scanning spectrometer/monochromator of the Hettrick–Underwood type with a constant deviation angle and a nearly constant focal distance at wavelengths $\lambda < 30$ nm; (iii) a flat-field 0.5-m long spectrograph of the Harada type. Broadband (11–14 nm) aperiodic Mo/Be multilayers were designed for extending the range of the imaging spectrograph to 11 nm. Numerical techniques were developed to calculate interference lithography configurations that provide the desired line density variation across the aperture of plane/concave VLS gratings.

25.1 Introduction

Investigations of laboratory/astrophysical plasmas and laboratory soft X-ray (XUV) radiation sources invite the use of diversified high-resolution spectroscopic instrumentation. Some of the spectrometers should have the capacity to record XUV spectra simultaneously in a broad spectral range of over an octave in wavelength. Also required are scanning spectrometers/monochromators capable of studying in detail a

E. N. Ragozin (✉) · A. O. Kolesnikov · A. N. Shatokhin · E. A. Vishnyakov
P.N. Lebedev Physical Institute, Russian Academy of Sciences, 53 Leninsky prosp., 119991 Moscow, Russia
e-mail: enragozin@gmail.com

A. O. Kolesnikov · A. N. Shatokhin
Moscow Institute of Physics and Technology (National Research University), 9 Institutskiy per., 141701 Dolgoprudnyi, Moscow Region, Russia

A. S. Pirozhkov
Kansai Photon Science Institute, National Institutes for Quantum and Radiological Science and Technology, Kizugawa, Kyoto 6190215, Japan

P. V. Sasorov
Institute of Physics ASCR, v.v.i. (FZU), ELI-Beamlines Project, 18221 Prague, Czech Republic

M. Kozlová and J. Nejdl (eds.), *X-Ray Lasers 2018*, Springer Proceedings in Physics 241, https://doi.org/10.1007/978-3-030-35453-4_25

narrow portion of the spectrum selected from the broad operating range. Some objects call for the use of stigmatic (imaging) spectrometers to record space-resolved spectra, sometimes with a spatial resolution on a micrometer scale. Among the recently emerged laboratory objects that invite the use of imaging (stigmatic) instruments are the source of high-order (even and odd) harmonics in a relativistic helium plasma produced by a multiterawatt femtosecond laser [1] and a relativistic 'flying mirror' [2]. A requirement common of all these spectrometers is that the diffracted radiation be normally incident on the focal surface to make them compatible with modern CCD detectors. This requirement is met by varied line-space (VLS)-grating spectrometers. Here we report the design and implementation of three VLS-grating spectrometers, which were tested by recording the line spectra of laser-produced plasmas. The plane and concave VLS gratings used in the spectrometers were made by interference lithography. A code package was developed to calculate the optical configuration for making plane and concave VLS gratings by interference lithography.

25.2 Stigmatic Spectrograph

The optical design of the inherently stigmatic (imaging) spectrometer was comprehensively described elsewhere [3]. The high-resolution stigmatic spectrograph makes combined use of a near-normal incidence focusing (spherical) multilayer mirror (MM) and a plane grazing-incidence VLS grating with the central line density 600 mm^{-1}. The instrument offers a combination of properties that was previously inherent in spectrometers of the visible and UV ranges: practical stigmatism throughout a broad spectral range, a high (practical) spectral resolving power (~1000), a relatively large solid acceptance angle $(2.9 \cdot 10^{-4}$ sr), and a large field of view (1 cm). The average plate scale is equal to 0.6 nm/mm. The operating range is defined by the spectral reflectivity curve of the MM. For the first-version instrument, we used a mirror of radius $R = 1$ m coated with an aperiodic Mo/Si multilayer structure specifically designed for maximum uniform reflectivity in the 12.5–25 nm range [4]. The MM was synthesized by magnetron sputtering in the Laboratory directed by V. V. Kondratenko in the National Technical University "Kharkov Polytechnic Institute," Kharkov, Ukraine. In the 12.5–25 nm range, the measured MM reflectivity varies between 18 and 13% and extends to at least 30 nm (~8%). In the future, the operating range will be extended down to 11 nm using an aperiodic Mo/Be MM (see below). The plane VLS grating and its fabrication are described in Sect. 25.4. The practical spectral resolution of the spectrograph is limited by the detector resolution and is numerically equal to the product of the plate scale and the size of two detector pixels $(2 \times 13$ μm). The spatial resolution in the vertical direction (parallel to the entrance slit) is also limited by detector resolution and is equal to 26 μm. The instrument measures 1.1 m in length.

By way of example Fig. 25.1 shows the line spectrum Mg-target plasma with spatial resolution in the direction of the target normal, which was recorded in one

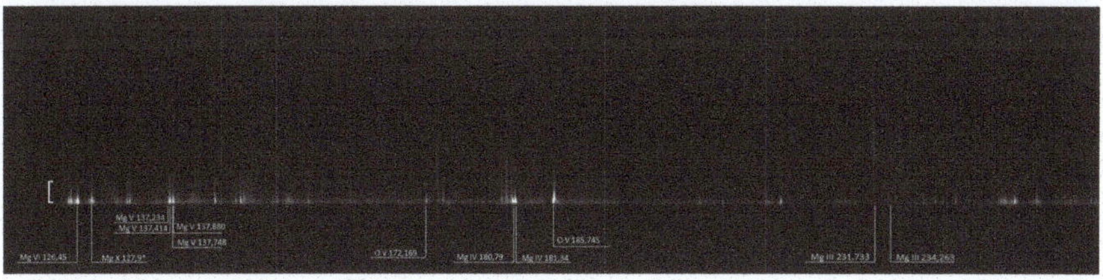

Fig. 25.1 Space-resolved spectrum of Mg plasma obtained in one laser shot. The bracket at the left indicates a 0.5 mm vertical scale (Borrowed from [3])

laser shot (0.5 J, 8 ns, 1.06 μm). A sharp light–shadow boundary testifies to a high spatial resolution (two detector pixels, 26 μm).

The operating range will be extended to 11 nm with the use of an aperiodic Mo/Be MM. The multilayer-structure optimization code was upgraded so as to minimize the thickness variation of the neighbor layers of the same name and thereby facilitate the multilayer synthesis. Figure 25.2 shows the simulated reflectivity of a Mo/Be structure designed for maximum uniform reflectivity between 11.1 and 13.8 nm and for minimal thicknesses variation of the neighbor layers of the same name. Also shown in Fig. 25.2 are the layer thicknesses of the corresponding Mo/Be structure capped with a Ru layer. Beryllium-based MMs are now synthesized in the Institute for the Physics of Microstructures, Nizhny Novgorod, Russia.

A reflectivity level of ~15% throughout the specified range seems to be achievable with the aperiodic Mo/Be MMs (see Fig. 25.2), which is quite favorable to the implementation of the imaging spectrograph in the wavelength range bounded below by the K edge of beryllium at 11.1 nm.

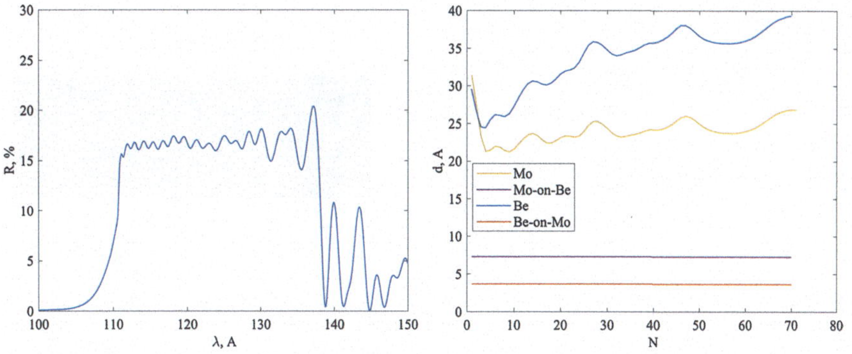

Fig. 25.2 Reflectivity of an aperiodic Mo/Be structure optimized for maximum uniform reflectivity in the 111–138 Å range with the inclusion of Mo-on-Be and Be-on-Mo transition layers (left), and its layer thicknesses (70 layer pairs, 3 nm Ru cap layer)

25.3 Scanning Spectrometer/Monochromator

A plane VLS grating with the same parameters as for the imaging spectrograph was used in the scanning spectrometer/monochromator of the Hettrick–Underwood type [5]. It comprises a spherical Au-coated mirror of radius $R = 6$ m mounted at a grazing angle of 8.34°, which produces the horizontal focus of the entrance slit at 533.7 mm behind the grazing-incidence VLS grating. The grating operates in the outside diffraction order, in the mode with a constant deviation angle of 16.68°, so that the overall deviation angle of the instrument is zero. Wavelength scanning is performed by grating rotation. The instrument may be used as a scanning flat-field spectrometer with a plate scale of ~0.3 nm/mm (on average) or, optionally, as a monochromator. The grating–detector distance (or the distance between the grating and the output slit) is fixed at 532.6 mm. The focal length is exactly equal to this distance at wavelengths of 14.0 and 27.3 nm. The focal length is nearly wavelength-independent, so that the geometric defocusing is less than 13.5 μm (the CCD detector pixel size) in the $\lambda < 33$ nm range. The short-wavelength bound of the operating range is determined by the source brightness rather than by the defocusing.

The scanning instrument was tested in the spectrometer mode by recording the line spectra of laser-produced plasma of LiF, Mg, and polyethylene targets. Figure 25.3 shows a portion of LiF plasma spectrum in a range of 120–150 Å. In the carbon ion spectrum, it was possible to resolve the fine structure (0.14 Å) of the H_α line (3 → 2, 182 Å), which testifies to a resolving power of 1300 at this wavelength.

The instrument was mounted on a 2-m long optical bench and accommodated in a 3.8-m long vacuum chamber 0.9 m in diameter, which was evacuated to below 10^{-4} Torr. The function of detector was fulfilled by a Greateyes backside-illuminated in-vacuum UV1 camera (2048 × 512 pixels of size 13.5 μm).

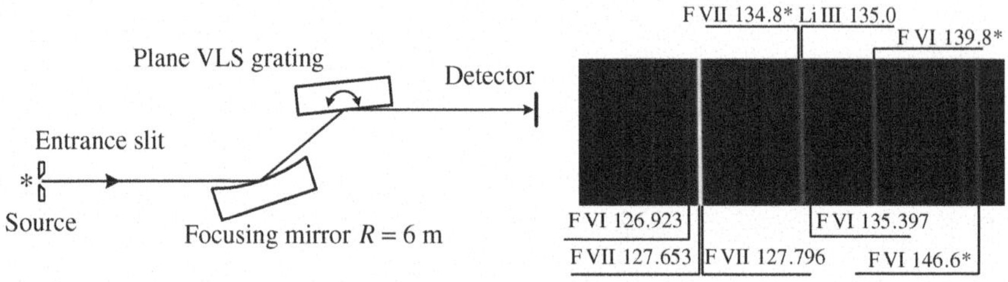

Fig. 25.3 Schematic of the scanning spectrometer/monochromator (left) and portion of LiF plasma spectrum recorded in the scanning spectrometer mode for one (fixed) VLS-grating rotation angle with a 20-μm wide entrance slit. The figures indicate wavelengths in Angstroms. Asterisks indicate unresolved line groups

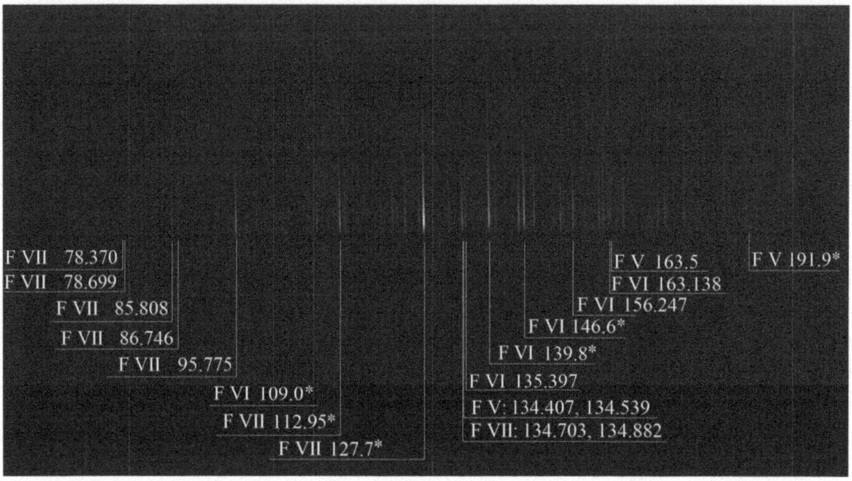

Fig. 25.4 Spectrum of a Teflon target recorded with our flat-field spectrograph with the use of 0.5-J, 8-ns, 1.06-μm laser pulses. The half-height linewidths of isolated lines correspond to two detector pixels (26 μm)

25.4 Flat-Field Spectrograph for a Range of 5–7.5 nm

The flat-field spectrograph of Harada, class comprises only one optical element: a spherical VLS grating. In [6], we designed three flat-field spectrographs of length 0.25, 0.5, and 1.5 m intended for operation in the respective ranges 9–25, 5–27.5, and 2–11 nm. For the first implementation, we selected the 0.5-m instrument with a grating of radius 6000 mm with a central line density of 1200 mm^{-1}. It operates in the inside orders of diffraction. The plate scale ranges from ~0.42 nm/mm at 6 to ~0.72 nm/mm at 22 nm. The resolving power corresponding to two detector pixels is wavelength-dependent and varies from 550 at a wavelength of 6 nm to 1170 at 22 nm. The instrument was used to record the line spectra from laser-produced plasmas, which amply confirmed our expectations. Figure 25.4 shows the line spectrum recorded from a Teflon target.

25.5 VLS-Grating Fabrication Technology

The VLS gratings used in the three spectrometers described above were made by interference lithography. At the heart of this technology is the design of the optical configuration that provides the requisite interference fringe density variation across the aperture of a photoresist-coated substrate. This dependence is conveniently expressed with a polynomial

$$p(w) = p_0 + p_1 w + p_2 w^2 + p_3 w^3, \quad (25.1)$$

where p_0 is the central line density. As is well known, factor p_1 modifies the spectral focal curve, while p_2 and p_3 may be used to cancel meridional coma and spherical aberration at fixed wavelengths. Namioka and Koike [7] solved the direct problem and derived explicit expressions for p_i for a given optical configuration with one (or two) auxiliary spherical (ellipsoidal) mirrors.

We adopted an optical configuration with one auxiliary spherical mirror and solved the inverse problem of interference lithography. We now have the capacity to design optical configurations for making ("writing") VLS gratings by interference lithography. In this case, the spatial interference fringe density variation across the aperture of a substrate satisfies the grating line density variation (25.1) required to achieve the desired spectral and spatial imaging. The last stage of the numerical procedure consists of the optimization of the configuration to lengthen the radius of line curvature with retention of the spectral resolution.

The superpolished substrates for our plane (55 × 25 mm) and spherical (Ø60 mm) gratings were made by SOTVEKS Ltd, Moscow. The lithographic procedure was performed in the State Institute of Applied Optics, Kazan, using 0.53-μm laser light. The gratings were next coated with 40 nm of gold.

Acknowledgements This work was supported by the Russian Science Foundation (Grant no. 14-12-00506).

References

1. A.S. Pirozhkov, M. Kando, T.Zh. Esirkepov et al., Burst intensification by singularity emitting radiation in multi-stream flows. Sci. Rep. (7), 17968 (2017)
2. M. Kando, A.S. Pirozhkov, K. Kawase et al., Enhancement of photon number reflected by the relativistic flying mirror. Phys. Rev. Lett. **103**, 235003 (2009)
3. A.N. Shatokhin, A.O. Kolesnikov, P.V. Sasorov et al., High-resolution stigmatic spectrograph for a wavelength range of 12.5–30 nm. Opt. Express **26**(15), 19009 (2018)
4. A.S. Pirozhkov, E.N. Ragozin, Aperiodic multilayer structures in soft X-ray optics. Phys. Usp. **58**(11), 1095 (2015) [Usp. Fiz. Nauk **185**(11), 1203 (2015)]
5. M.C. Hettrick, J.H. Underwood, Varied-space grazing incidence gratings in high resolution scanning spectrometers. AIP Conf. Proc. **147**, 237–245 (1986)
6. E.A. Vishnyakov, A.N. Shatokhin, E.N. Ragozin, Conception of broadband stigmatic high-resolution spectrometers for the soft X-ray range. Quantum Electron. **45**(4), 371–376 (2015)
7. T. Namioka, M. Koike, Aspheric wave-front recording optics for holographic gratings. Appl. Opt. **34**(13), 2180–2186 (1995)

Chapter 26
Spectral and Polarization Properties of VUV-Mirrors for Experiments at a HHG Beamline

S. Richter, S. Espinoza and J. Andreasson

Abstract For polarization-resolved reflection experiments such as ellipsometry, not only is high reflectivity in a wide spectral range required for mirrors but also the quality of their polarization response is important. Furthermore, for VUV ellipsometry, optimal angles of incidence at the sample are between 45° and 60° with respect to the surface normal. In the best case, a setup should even allow variable angles. This requires reflective optics working at non-grazing incidence. In this theoretical study, a selection of potentially relevant materials and mirror designs for broadband use in the VUV is investigated. Based on the available tabulated databases of optical constants, we performed transfer-matrix calculations to obtain reflectance as well as polarization-response spectra in the desired VUV range up to 50 eV. From the variety of materials, we discuss metals that are otherwise commonly used at grazing incidence, Si, SiC, and as representatives for coating layers MgF_2, SiO_2, and Al_2O_3. While SiC is most universal, Si with only native oxide layer performs well especially below 25 eV. Aluminum has good properties but oxide layers are very detrimental, as protective coatings are in general.

26.1 Introduction

The intermediate spectral range between ultraviolet (UV) and soft X-ray radiation is covered by synchrotrons and free-electron lasers but also laser-driven high-harmonic-generation (HHG) sources which allow generation of ultrashort VUV pulses for time-resolved experiments [1]. Here, the focus is on spectroscopy in the vacuum ultraviolet (VUV) range below 50 eV photon energy. Thus, broadband mirrors are required and Bragg mirror designs with rather narrow reflection bands do not suffice.

S. Richter (✉) · S. Espinoza · J. Andreasson
ELI Beamlines/Fyzikální ústav AV ČR, v.v.i., Za Radnicí 835,
25241 Dolní Břežany, Czech Republic
e-mail: steffen.richter@eli-beams.eu

J. Andreasson
Chalmers tekniska högskola, Institutionen för fysik, Kemigården 1,
41296 Göteborg, Sweden

M. Kozlová and J. Nejdl (eds.), *X-Ray Lasers 2018*, Springer Proceedings in Physics 241, https://doi.org/10.1007/978-3-030-35453-4_26

A further complication comes with polarization experiments like ellipsometry [2–4]. Here, not only high reflectance but also the polarization response of a mirror matters. Moreover, experimental configurations might require reflectance at non-grazing angles of incidence (AOI). This sets high demands on mirrors for VUV ellipsometry setups.

At a given AOI, a mirror (or any sample) is characterized by complex reflection coefficients $\mathbf{r}_{s/p}$ which depend on the photon energy E and the light polarization. The indices s/p refer to s (*senkrecht*/transverse electric, TE) and p (*parallel*/transverse magnetic, TM) polarization which are the linearly polarized basis states defined through the plane of incidence at the mirror. In ellipsometry, the ratio $\rho = \mathbf{r}_p/\mathbf{r}_s = \tan(\Psi)e^{i\Delta}$ is investigated [5]. If the experimental configuration requires additional mirrors, they are required to reflect both basis polarizations well at their set AOI, i.e., Ψ of the mirror should be close to 45°. Otherwise, a mirror would act as a polarizer. Furthermore, it is preferable if they do not introduce an additional phase shift which transfers (certain) linearly polarized light to be circularly polarized and vice versa, hence $\Delta \equiv 0° \bmod 180°$. Still, having such a polarization response of the mirrors in use is only practical if the reflectance $R_{s/p} = |\mathbf{r}_{s/p}|^2$ is sufficiently large. Hence, these are the quantities to be investigated regarding good compromises.

26.2 VUV Optical Constants

In the spectral transition range between optical and X-ray radiation, a cross-over of different conventions appears. We represent the complex refractive index $n + i\kappa$ with refractive index $n = 1 - \delta$ and extinction coefficient $\kappa = \beta$, which is related to the dielectric function $\varepsilon_1 + i\varepsilon_2$ as $n + i\kappa = 1 - \delta + i\beta = \sqrt{\varepsilon_1 + i\varepsilon_2}$. The parametrization by δ and β is most common convention toward X-ray optics [6].

The most comprehensive source for VUV optical constants (partially measured by VUV ellipsometry at the synchrotron) is Palik's book [7]. A good selection can also be found in Weber's book [8]. From 50 eV (to 30 keV) the Henke database [6] provides optical constants computed from semi-empirical atomic scattering factors. While giving good estimates, these computed optical constants can be problematic especially in the lower energy region [7]. Figure 26.1 shows spectra of the complex refractive index for various materials potentially relevant in the field of VUV optics. It should be noted that there are hardly any materials that do not absorb strongly for high photon energies. Above 30 eV, the refractive indices approach 1 for nearly all materials. Hence, reflectivity is mainly determined by absorption in that spectral range. Critical for a well-performing mirror is especially the spectral region below 25 eV because the optical response is strongly polarizing (see Figs. 26.2 and 26.3).

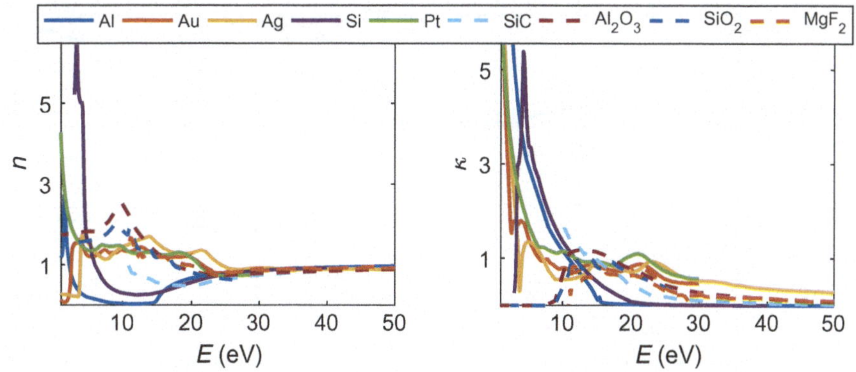

Fig. 26.1 Refractive index n and extinction coefficient κ of some relevant materials for the VUV spectral range

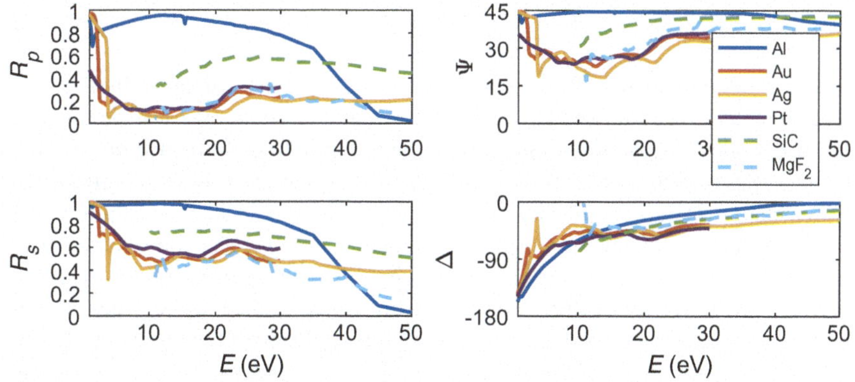

Fig. 26.2 Computed bulk-mirror properties. $R_p = |\mathbf{r}_p|^2$ and $R_s = |\mathbf{r}_s|^2$ are reflectance in p and s polarization, respectively. $\Psi = |\mathbf{r}_p/\mathbf{r}_s|$ and $\Delta = \mathrm{Arg}(\mathbf{r}_p/\mathbf{r}_s)$ are the ellipsometric angles expressing the polarization response. Preferable values are $\Psi \approx 45°$ and $\Delta \approx 0°$

26.3 VUV Mirrors

In order to investigate different materials and designs to be useful as VUV mirrors in setups for polarization experiments we performed transfer matrix calculations with full polarization treatment [9, 10]. As AOI we chose 70° measured from the surface normal. This relates to a particular goniometer configuration with a straight path between source and detector [4], allowing for an AOI at the sample of 50° which is advantageous for ellipsometry at such high photon energies [2].

The reflectivity of representative materials as bulk mirrors with a perfect surface is shown in Fig. 26.2. Generally, it can be noted that the polarization performance improves for higher energies while reflectivity drops. Though commonly used for grazing-angle optics at synchrotrons, gold, silver, and platinum are not a good choice

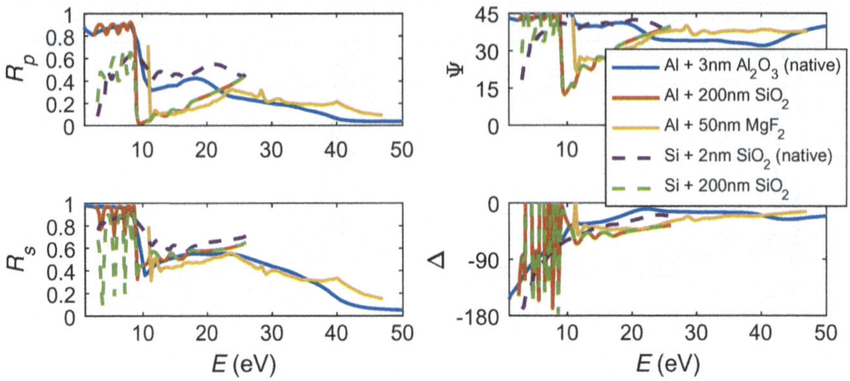

Fig. 26.3 Computed performance of coated mirrors of different materials according to the legend. Shown quantities are similar to Fig. 26.2. Interferences from the coating layers appear in the spectral range below 10 eV where the materials are transparent

for non-grazing incidence. As an alternative to those metals, silicon carbide reveals an acceptable level of reflectivity and behaves quite constant over a wide range of photon energies. Aluminum seems to have an unbeatable performance up to 40 eV. However, practically it is impossible to keep a clean surface without a native oxide layer. This is why a protective coating must be applied to avoid oxidization, which is often done. Various mirror designs with single (coating) layers on top have been computed; a selection is demonstrated in Fig. 26.3. The figure compares response curves of bulk mirrors with thin natural oxide layers (3 nm Al_2O_3 on Al, 2 nm SiO_2 on Si) with common protective coatings (50 nm MgF_2, 200 nm SiO_2). From comparison with Fig. 26.2, it is obvious that any thicker oxide or coating dominates the response, no matter what material is below. Furthermore, any oxide or coating seems to be detrimental to mirror performance. This is due to the high absorption of the coating materials for high photon energies. Surprisingly, silicon with only native oxide (which remains stable) performs relatively well in the rather problematic spectral range below 25 eV. Other designs based on more materials have been computed but were not convincing for use as mirrors.

In addition, the effect of roughness has been investigated by different means of non-idealities: Roughness with characteristic structure size smaller than the wavelength of light can be described by effective-medium theory as a mixture of the material and void. Such roughness turns out to have relatively minor effects, mainly decreasing the *s*-polarized reflectivity for high photon energies. If, however, roughness structures are larger than the wavelength of light, the rough surface can cause depolarization. Computations confirm that this is only relevant if coating layers are transparent, hence it seems hardly relevant for higher photon energies (cf. Fig. 26.1) in the context of ellipsometry experiments.

26.4 Conclusion

For non-grazing angles of incidence, the material choice for broadband VUV mirrors up to 50 eV is challenging. By computing different mirror designs based on available database optical-constants, it can be concluded that

- Silicon carbide has acceptable reflectance and the most constant performance over a wide spectral range.
- Metals are generally not convincing, except Al if it would be available without oxide layer.
- SiO_2 and Al_2O_3 are very disadvantageous, MgF_2 can be acceptable. Respective protective coatings should be as thin as possible.
- Silicon with only a thin, native oxide layer appears to be a cheap, sustainable alternative especially for photon energies up to 25 eV.

It should also be mentioned that some surfaces or their combination are potentially applicable as reflective polarizers or retarders.

Acknowledgements This work was supported by the projects *Advanced research using high-intensity laser produced photons and particles* (*ADONIS*) (CZ.02.1.01/0.0/0.0/16_019/0000789) and *Structural dynamics of biomolecular systems* (*ELIBIO*) (CZ.02.1.01/0.0/0.0/15_003/ 0000447) from the European Regional Development Fund. S. Richter was supported by the Czech Institute of Physics Mobility project (CZ.02.2.69/0.0/0.0/16_027/0008215). J. Andreasson acknowledges the support of the Ministry of Education, Youth and Sports as part of targeted support from the National Programme of Sustainability II and the Chalmers Area of Advanced Materials Science.

References

1. O. Hort et al., Opt. Express **27**, 8871–8883 (2019)
2. R.L.M. Johnson et al., Rev. Sci. Instrum. **60**, 2209 (1989)
3. M.D. Neumann et al., Rev. Sci. Instrum. **85**, 055117 (2014)
4. S. Espinoza et al., Appl. Surf. Sci. **421**, 378–382 (2017)
5. H.G. Tompkins, E.A. Irene, *Handbook of Ellipsometry* (Springer, 2005)
6. B.L. Henke et al., Atom. Data Nucl. Data Tables **54**(2), 181–342 (1993) [henke.lbl.gov]
7. E.D. Palik, *Handbook of Optical Constants of Solids* (Academic Press, 1998)
8. M.J. Weber, *Handbook of Optical Materials* (CRC Press, 2003)
9. D.W. Berreman, JOSA **62**(4), 502–510 (1972)
10. M. Schubert, Phys. Rev. B **53**(8), 4265 (1996)

Chapter 27
On Inverse Problem for Propagation of Waves from Inclined Surfaces

R. M. Feshchenko

Abstract The inverse problem is considered in the framework of 2D parabolic wave equation with the initial conditions specified on an inclined line. It is demonstrated that the inversion can be performed by the solution of one singular integral equation of Cauchy type, which in turn can be transformed into a system of linear algebraic equations and solved numerically. The existence, uniqueness and stability of numerical solutions are studied. Several model solutions are considered as examples.

27.1 Introduction

The parabolic wave equation (PWE), which was first introduced by Leontovich and Fock more than 60 years ago [1], is widely used in computational physics and engineering to describe the propagation of paraxial or quasi-paraxial electromagnetic [2] and acoustics [3] beams in free space as well as in inhomogeneous media. It was successfully applied for solution of complex problems in laser physics [4], electromagnetic radiation propagation [2], underwater acoustics [3, 5], X-ray optics [6], microscopy, and lenseless imaging [7, 8].

One of the underutilized mathematical properties of the PWE is a possibility to express the field amplitude in a part of free space through the initial values of the amplitude specified on an inclined line or plane. However, the inverse problem in this case has not been thoroughly considered although we obtained an integral equation for its solution in the case of 2D PWE in one of our previous works [9].

In this paper, we analyse the existence and uniqueness of solutions of this inclined parabolic inverse problem (IPIP) and perform numerical experiments for a number of simple initial amplitude models to test the stability and convergence of its solutions.

R. M. Feshchenko (✉)
P.N. Lebedev Physical Institute of RAS, 53 Leninski Prospect,
Moscow 119991, Russia
e-mail: rusl@sci.lebedev.ru

M. Kozlová and J. Nejdl (eds.), *X-Ray Lasers 2018*, Springer Proceedings in Physics 241, https://doi.org/10.1007/978-3-030-35453-4_27

27.2 Direct and Inverse Problem Solution

Let us consider the 2D parabolic wave equation [2]

$$2ik\frac{\partial u}{\partial z} + \frac{\partial^2 u}{\partial x^2} = 0, \tag{27.1}$$

when the initial wave field amplitude u_0 is specified on an inclined line defined as $x + \tan\theta z = 0$, $u_0(z) = u(\tan\theta z, z)$, $z < 0$ where θ is the angle between this inclined line and axis z, $k = 2\pi/\lambda$ is the wave number, z is the coordinate along the beam propagation direction and x is the transversal coordinate. Then the field amplitude in the domain $x > -z\tan\theta$ can be expressed as [9]

$$u(x,z) = (x\cos\theta + z\sin\theta)\times \sqrt{\frac{k}{2\pi i}} \int_{-z/\cos\theta}^{\infty} \frac{u_0(-s\cos\theta)}{(z + s\cos\theta)^{3/2}} \exp\left[\frac{ik(x - s\sin\theta)^2}{2(z + s\cos\theta)}\right] \mathrm{d}s. \tag{27.2}$$

Let us assume that the field amplitude $u(x,z)$ is known on the semi-line $x \geq 0$, $z = 0$. The inverse problem is to find the field amplitude u at the semi-infinite inclined line defined above. The integral equation for the simplified case when $\theta = 0$, which was obtained earlier [9], is

$$u_0(z'') - \frac{i}{\pi}\sqrt{z - z''}\, P\int_{-\infty}^{z} \frac{u_0(\zeta)}{z'' - \zeta} \frac{\mathrm{d}\zeta}{\sqrt{z - \zeta}} = \sqrt{\frac{2ki}{\pi(z - z'')}} \int_0^{\infty} u(x', z) \exp\left[-\frac{ikx'^2}{2(z - z'')}\right] \mathrm{d}x', \tag{27.3}$$

where the integral in the second term on the left side of (27.3) is understood as the Cauchy principal value.

Equation (27.3) can be further simplified by assuming $z = 0$ and introducing a new integration variable $\mu = -\zeta$, a new independent variable $t = -z'' > 0$ and a new function $v(t) = u_0(-t)/\sqrt{t}$, which we will also call the amplitude. Then (27.3) can be rewritten as

$$v(t) + \frac{1}{\pi i} P\int_0^{\infty} \frac{v(\mu)}{\mu - t} \mathrm{d}\mu = \sqrt{\frac{2ki}{\pi}} \frac{1}{t} \int_0^{\infty} u(x', 0) \exp\left[-\frac{ikx'^2}{2t}\right] \mathrm{d}x' = H(t), \tag{27.4}$$

which is a singular integral equation of Cauchy type [10].

From (27.4), it directly follows that in two special cases—when the field amplitude is a priori known to be either a real or imaginary function, the inverse problem can be solved exactly by applying complex conjugation to (27.4) and adding or subtracting the result from the initial equation. For instance, in the real case this gives us the following expression for the field amplitude at the semi-line $x = 0,\ z < 0$

$$v(t) = \frac{1}{2}\left[H(t) + H^*(t)\right] = \operatorname{Re} H(t). \tag{27.5}$$

Let us consider variable t to be a complex number. Then the sum of the principle value integral and the amplitude value at point t in (27.4) can be combined into one complex path integral where the path is taken around the pole from the below. Then (27.4) will take the following form:

$$\frac{1}{\pi i} \int_0^{\infty} \frac{v(\mu)}{\mu - t}\, \mathrm{d}\mu = H(t). \tag{27.6}$$

From the general theory of Cauchy singular equations it is known (see [10]) that a solution of (27.6) exists if and only if its right side $H(t)$ is a sectionally analytical function with a branch cut along the positive semi-axis. In addition, if a solution of (27.6) exists it is necessary unique [10]. From the same theory it is also known that the solution, if it exists, is equal to the jump of function $H(t)$ across its branching cut [10]

$$v(t) = H^+(t) - H^-(t), \tag{27.7}$$

where $H^{\pm}(t)$ are the values of $H(t)$ on the upper/lower sides of the cut. Therefore, if an analytical expression for function $H(t)$ is known, finding the solution or proving that it does not exist is a trivial task although in practice it is not that simple.

27.3 Numerical Experiments

Let us assume that the field amplitude in the image plane at $z = 0$ is known in $N_x + 1$ points $\{x_i\}$, where $i = 0, \ldots N_x$, $x_i - x_{i-1} = h$ and that outside the interval $[x_0, x_{N_x}]$ the amplitude is zero. Then the right-hand side of equation (27.4) can be approximated using an appropriate summation (for instance, trapezoidal) rule.

To approximate the principle value integral in the left-hand side of (27.4) one can assume that initial amplitude u_0 deviates from zero only in an interval of z: $z_{\max} \geq z \geq z_{\min}$, which is then split into N parts and the value of the integral at $\{z_n\}$ is approximated as a sum of integrals over intervals $[z_m, z_{m+1}]$, where $n, m = 0, \ldots N$, $z_m = z_{\min} + \tau m$ and $\tau = (z_{\max} - z_{\min})/N$.

Eventually, the following system of linear algebraic equation can be obtained:

$$\left(\mathbf{I} - \frac{\mathbf{M}}{\pi i}\right) \overrightarrow{\mathbf{u_0}}^T = \overrightarrow{\mathbf{g}}^T, \tag{27.8}$$

where $\mathbf{M}$ is a matrix $m \times n$, $\mathbf{I}$ is a unity matrix, $\overrightarrow{\mathbf{u_0}}$ is a row vector made of values $u_0(z_n)$ and $\overrightarrow{\mathbf{g}}$ is a row vector made of values g_n of the approximated right hand side integral in (27.4). The expressions for elements of matrix $\mathbf{M}$ and vector $\overrightarrow{\mathbf{g}}$ can be found in [11].

Let us now demonstrate practical utilization of (27.4) by a number of numerical experiments involving a Gaussian beam, a parabolic beam and a step-like initial amplitude $u_0(z)$ (see details in [11]). In Fig. 27.1a–c the squared modules of the pre-specified initial amplitudes are shown together with their reconstructed values obtained using linear system (27.8). It can be seen that the reconstructed amplitudes for the Gaussian and parabolic beams closely approximate the respective pre-specified initial values although even in all three models the matrices $\mathbf{M}$ are near singular and their condition numbers are high—about $8 \cdot 10^{12}$ (matrix $\mathbf{M}$ does not depend on the model used). However in the case of step-like initial amplitude, there exists a significant difference with the pre-specified initial amplitude especially near the borders of the computational domain $[z_{\min}, z_{\max}]$.

The general stability/instability of numerical solutions is further illustrated in Fig. 27.1d where the square difference function $D(\tau) = \sum_{n=0}^{N} |u_0^1(z_n) - u_0^2(z_n)|^2 / |u_0^1(z_n)|^2$ is plotted against the longitudinal step τ for the solutions of (27.8) ($u_0^2(z_n)$) and the initial amplitude ($u_0^1(z_n)$).

It should be noted that in Fig. 27.1 we chose the computational domain $[z_{\min}, z_{\max}]$ to be wider than the interval where the step-like amplitude deviates from zero, which made the solution noticeably better. However, it is clear that step-like initial profiles may still not be suitable for the reconstruction.

27.4 Conclusions

Using the parabolic approximation we solved IPIP by reversing the expression for the field amplitude propagating from an inclined line. We demonstrated that IPIP is reduced to solution of a singular Cauchy-type integral equation in a semi-infinite domain. The necessary and sufficient conditions for the existence of solutions of this integral equation were determined and it was shown that if a solution exists it is necessarily unique.

We demonstrated that IPIP can be efficiently solved numerically by reducing the obtained singular integral equation to a system of linear algebraic equations. We conducted several numerical experiments using a Gaussian beam, a parabolic-like initial amplitude and a step-like initial amplitude as examples. We demonstrated that for smooth initial field amplitudes of Gaussian and parabolic beams vanishing at the computational domain borders a good numerical solution of IPIP can be obtained by appropriate selection of the longitudinal step, the image plane integration interval and other parameters.

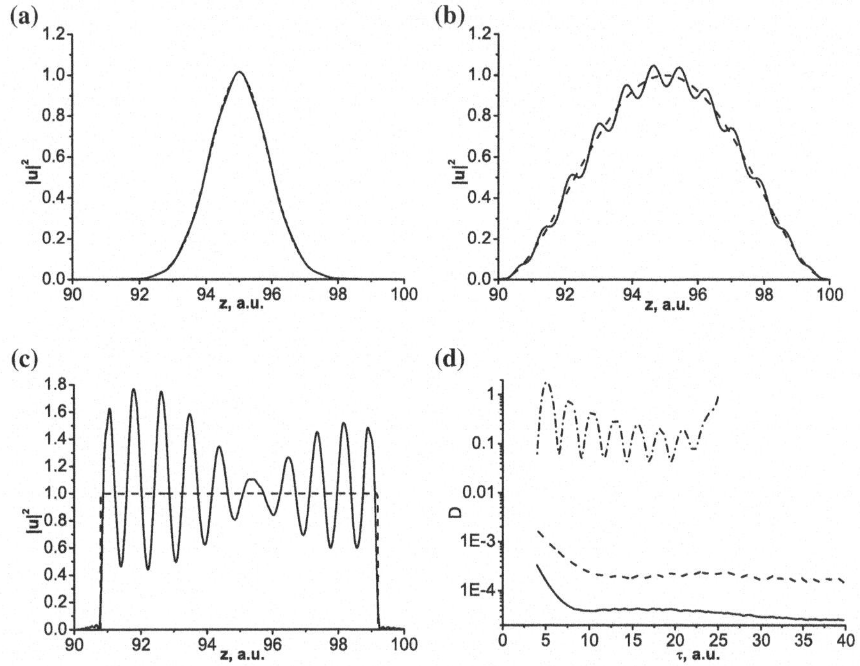

Fig. 27.1 The squared module of the initial field amplitude (**a**–**c**) as function of z and difference function $D(\tau)$ (**d**) as function of the longitudinal step τ for three model fields: Gaussian beam—**a** (solid curve in **d**), parabolic beam—**b** (dash curve in **d**) and step like profile—**c** (dash-dot curve in **d**). In (**a**–**c**) dash lines are initial amplitudes and solid lines are reconstructions using (27.8). The following parameters were used: $z_{\min} = 90$, $z_{\max} = 100$, $\tau = 0.00398$, $\lambda = 0.01$, $\Lambda = 1$ (the period of oscillations), $x_{\min} = 2.8284$, $x_{\max} = 28.284$, $N_x = 2828$, $N = 2514$, $h = 0.01$, $L = 20$ (Rayleigh length of the Gaussian beam) and $a = 4.8$ (semi-width of the parabolic and step-like profiles)

Acknowledgements The work was supported by the Basic Research Programme of the Presidium of the Russian Academy of Sciences *Actual problems of photonics, probing inhomogeneous mediums and materials*.

References

1. V.A. Fock, *Electromagnetic Diffraction and Propagation Problems* (Pergamon Press, Oxford, 1965)
2. M. Levy, *Parabolic Equation Methods for Electromagnetic Wave Propagation*. IEE Publication Series (Institution of Electrical Engineers, 2000)
3. D. Lee, A.D. Pierce, J. Comput. Acoust. **3**(02), 95 (1995)
4. M.S. Sodha, A.K. Ghatak, V.K. Tripathi, Prog. Opt. **13**, 169 (1976)
5. M. Spivack, J. Acoust. Soc. Am. **95**(2), 694 (1994)
6. A. Sakdinawat, D. Attwood, Nat. Photonics **4**(12), 840 (2010)

7. H.N. Chapman, K.A. Nugent, Nat. Photonics **4**(12), 833 (2010)
8. P. Thibault, V. Elser, Annu. Rev. Condens. Matter Phys. **1**(1), 237 (2010)
9. I.A. Artyukov, R.M. Feshchenko, N.L. Popov, A.V. Vinogradov, J. Opt. **16**(3), 035703 (2014)
10. R. Estrada, R.P. Kanwal, *Singular Integral Equations* (Springer Science & Business Media, New York, 2012), pp. 71–116
11. R.M. Feshchenko, On inverse problem for an inclined line in the parabolic approximation (2018), arXiv:1804.04718

Chapter 28
Flux and Energy Density of a Collapsing Electromagnetic Pulse in the Free Space

I. A. Artyukov, A. V. Vinogradov, N. V. Dyachkov and R. M. Feshchenko

Abstract On the basis of the exact solution of the Maxwell equations, we obtained a relationship between the energy and spectrum of a collapsing electromagnetic wave and the energy density in the focusing point. The presented theoretical approach can be useful for a study of ultra short laser pulses without approximation of slowly varying amplitudes.

28.1 Introduction

How much electric field intensity can be obtained on focusing of a laser pulse with given energy? Let us answer this question by using one of exact solutions of the Maxwell equation in the free space:

$$\nabla \times \vec{E} = -\frac{1}{c}\frac{\partial \vec{H}}{\partial t}, \tag{28.1}$$

$$\nabla \times \vec{H} = \frac{1}{c}\frac{\partial \vec{E}}{\partial t}, \tag{28.2}$$

$$\nabla \cdot \vec{E} = 0, \tag{28.3}$$

$$\nabla \cdot \vec{H} = 0. \tag{28.4}$$

In this case, the scalar potential can be chosen equal to 0 and (28.1)–(28.4) reduce to two equations for the vector potential $\vec{A}$:

I. A. Artyukov · A. V. Vinogradov (✉) · N. V. Dyachkov · R. M. Feshchenko
P.N. Lebedev Physical Institute RAS, Moscow, Russia
e-mail: vinograd@sci.lebedev.ru

N. V. Dyachkov
Moscow Institute of Physics and Technology (State University), Dolgoprudnyi, Moscow, Russia

M. Kozlová and J. Nejdl (eds.), *X-Ray Lasers 2018*, Springer Proceedings in Physics 241, https://doi.org/10.1007/978-3-030-35453-4_28

$$\Delta\vec{A} - \frac{1}{c^2}\frac{\partial^2\vec{A}}{\partial t^2} = 0, \quad \nabla \cdot \vec{A} = 0, \tag{28.5}$$

with

$$\vec{E} = -\frac{1}{c}\frac{\partial\vec{A}}{\partial t}, \quad \vec{H} = \nabla \times \vec{A}. \tag{28.6}$$

One can see that the following form of the potential satisfies the (28.5):

$$\vec{A} = \vec{l} \times \nabla u, \tag{28.7}$$

where $\vec{l}$ is the axial unit vector which is independent of time and coordinate and $u(\vec{r}, t)$ satisfies the (28.5). The spherically symmetrical solution $u(r, t)$ consists of two counter-propagating spherical waves described by arbitrary functions f_+ and f_-:

$$u(r, t) = \frac{f_+(r - ct) + f_-(r + ct)}{r}, \tag{28.8}$$

To avoid a singularity at the center $r = 0$ the solution can be written as [1]:

$$u(r, t) = \frac{f(ct + r) - f(ct - r)}{r}, \tag{28.9}$$

where $f(x)$ is an arbitrary function as well with $-\infty < x < \infty$. For a function rapidly decreasing for large values of $|x|$, the energy of electromagnetic pulse (or beam) is limited.

Using (28.6) and vector potential (28.7), (28.9) one can obtain expressions for the fields $\vec{E}$ and $\vec{H}$:

$$\vec{E} = -\frac{1}{c}\frac{\partial\vec{A}}{\partial t} = -\frac{1}{c}\frac{\partial^2 u}{\partial t\partial r}\vec{l} \times \vec{n}, \tag{28.10}$$

$$\vec{H} = \nabla \times \vec{A} = \left(\Delta u - \frac{1}{r}\frac{\partial u}{\partial r}\right) \cdot \vec{l} - \left[r\frac{\partial}{\partial r}\frac{1}{r}\frac{\partial u}{\partial r}\right]\left(\vec{l} \cdot \vec{n}\right) \cdot \vec{n}, \tag{28.11}$$

where $\Delta u = \frac{1}{r^2}\frac{\partial}{\partial r}\left(r^2\frac{\partial u}{\partial r}\right)$, $\vec{n}$ is the unit vector along a direction of $\vec{r}$. Formulas (28.9)–(28.11) contain an arbitrary function f and describe some sort of light beams of limited energy, which satisfy the Maxwell equations in the free space. Thus, for a finite function $f(x)$ the pulse looks like a collapsing spherical shell. On increasing the time t from $-\infty$ to $+\infty$ it moves towards its center to collapse and then expands.

28.2 The Pulse Energy

The total energy ε of the pulse (28.9)–(28.11) can be found by the formula:

$$\varepsilon = \lim_{r\to\infty} \int_0^\infty \Phi(r,t)\mathrm{d}t, \tag{28.12}$$

where $\Phi(r,t)$ is the flux of energy through the sphere of radius r:

$$\Phi(r,t) = r^2 \iint \sin\theta \mathrm{d}\theta \mathrm{d}\varphi I(\vec{r},t), \tag{28.13a}$$

and $I(r,\theta,t)$ is the radial intensity

$$I(\vec{r},\theta,t) = \frac{c}{4\pi}\overrightarrow{n}\cdot\left[\vec{E}\times\vec{H}\right]. \tag{28.13b}$$

Now substituting the field strengths (28.10) and (28.11) into (28.13b) and (28.13a) we get

$$\Phi(r,t) = r^2\frac{2}{3}\frac{\partial^2 u}{\partial t\partial r}\left(\frac{1}{r}\frac{\partial u}{\partial r} - \Delta u\right). \tag{28.14}$$

Putting $\Delta u = \frac{1}{r^2}\frac{\partial}{\partial r}r^2\frac{\partial u}{\partial r}$ we can write:

$$\Phi(r,t) = -r^2\frac{2}{3}\frac{\partial v}{\partial t}\left(\frac{v}{r} + \frac{\partial v}{\partial r}\right), \quad v = \frac{\partial u}{\partial r}. \tag{28.15}$$

Then integrating the first term in (28.15) according to (28.12) we obtain:

$$\varepsilon = \lim_{r\to\infty} r^2\frac{2}{3}\left[\frac{v^2(r,t=0)}{2r} - \int_0^\infty \frac{\partial v}{\partial t}\frac{\partial v}{\partial r}\mathrm{d}t\right]. \tag{28.16}$$

$$\frac{\partial v}{\partial t} = \frac{c}{r}\left[f''(ct+r) + f''(ct-r)\right] - \frac{c}{r^2}\left[f'(ct+r) - f'(ct-r)\right]. \tag{28.17}$$

$$\frac{\partial v}{\partial r} = \frac{f''(ct+r) - f''(ct-r)}{r} - 2\frac{f'(ct+r) + f'(ct-r)}{r^2} + \frac{f'(ct+r) - f'(ct-r)}{r^2}. \tag{28.18}$$

In the product of expressions (28.17) and (28.18), we keep only higher orders of $\frac{1}{r}$:

$$\frac{\partial v}{\partial t}\frac{\partial v}{\partial r} = c\frac{\left[f''(ct+r)\right]^2 - \left[f''(ct-r)\right]^2}{r^2}. \tag{28.19}$$

Then after substitution of (28.19) into (28.16) we get

$$\varepsilon = \lim_{r\to\infty} \frac{2}{3}\left[-\int_0^\infty \frac{\partial v}{\partial t}\frac{\partial v}{\partial r}\mathrm{d}t\right] = \lim_{r\to\infty} \frac{2c}{3}\left[-\int_0^\infty [f''(ct+r)]^2\mathrm{d}t + \int_0^\infty [f''(ct-r)]^2\mathrm{d}t\right]$$

$$= \frac{2}{3}\int_{-\infty}^{\infty} [f''(x)]^2\mathrm{d}x. \tag{28.20}$$

Finally, the energy of the pulse (28.9)–(28.11) can be written as

$$\varepsilon = \frac{2}{3}\int_{-\infty}^{\infty} [g(x)]^2\mathrm{d}x,\ g(x) = f''(x). \tag{28.21}$$

28.3 Far-Field Spectrum of the Pulse

Principle formulas for the spectrum of a real signal function are given in Annex. To use them in our calculations we rewrite the formula (28.13b) for a radial intensity with axis Z along the vector $\vec{l}$. Then, substitution of the fields $\vec{E}$ and $\vec{H}$ from (28.9)–(28.11) leads to

$$I(\vec{r}, t) = \frac{\sin^2\theta}{4\pi}\left(\frac{1}{r}\frac{\partial u}{\partial r} - \Delta u\right)\frac{\partial^2 u}{\partial t \partial r}. \tag{28.22}$$

The fluence $F(\vec{r})$, i.e. radiant energy received by a surface per unit area, for large radius values r is

$$F(\vec{r}) = \lim_{r\to\infty}\int_0^\infty I(\vec{r}, t)\mathrm{d}t \tag{28.23}$$

Using approximate expressions for the fields $\vec{E}$ and $\vec{H}$ we can simplify formulas (28.12) and (28.23):

$$F(\vec{r}) = \frac{\sin^2\theta}{4\pi r^2}\int_{-\infty}^{\infty} [g(x)]^2\mathrm{d}t \tag{28.24}$$

This result looks to be reasonable because the numerator presents the total pulse energy ε. Now it is easy to derive the spectral density $F_\omega(\vec{r})$ (see (28.38) and (28.39) in Annex).

$$F_\omega(\vec{r}) = \frac{\sin^2\theta}{c^2r^2}|g(\kappa)|^2, \kappa = \frac{\omega}{c}, g(\kappa) = \frac{1}{2\pi}\int\limits_{-\infty}^{\infty} g(x)e^{i\kappa x}\mathrm{d}x, \tag{28.25}$$

$$F(\vec{r}) = \int\limits_0^\infty F_\omega(\vec{r})\mathrm{d}\omega.$$

Note that functions $g(x)$ and $|g(\kappa)|^2$ are suggested to be real and even, correspondingly. Evidently, the function $F_\omega(\vec{r})$ denotes the spectrum of a pulse in far-field region.

28.4 Energy Density in the Centre of Collapsing Pulse

We will determine the energy density in the center of pulse with using the formula

$$\epsilon = \left.\frac{\vec{E}^2 + \vec{H}^2}{8\pi}\right|_{r=0} \tag{28.26}$$

Before substitution of (28.10)–(28.11) into (28.26), we expand functions *u(r, t)* and *f(r, t)* into a Taylor series at $r = 0$:

$$\begin{aligned} f(ct+r) &= f(ct) + f'(ct)r + \tfrac{1}{2}f''(ct)r^2 + \tfrac{1}{6}f'''(ct)r^3 + \ldots \\ u(ct,r) &= 2f'(ct)r + \tfrac{1}{3}f'''(ct)r^2 + \ldots \\ u'(ct,r) &= \tfrac{2}{3}f'''(ct)r + \ldots \\ u''(ct,r) &= \tfrac{2}{3}f'''(ct) + \ldots . \end{aligned} \tag{28.27}$$

Then the expressions for electric and magnetic fields and the energy density in the center of collapsing pulse will be

$$\begin{aligned} \vec{E}(ct) &= 0, \vec{H}(ct) = \tfrac{4}{3}f'''(ct)\vec{l} \\ \epsilon(x) &= \tfrac{2}{9\pi}\left[g'(x)\right]^2, \quad x = ct. \end{aligned} \tag{28.28}$$

Equation (28.28) shows the EM energy density ϵ(t) at $r = 0$ increases with the increase of the time from t = $-\infty$ but later it goes down to 0.

28.5 Collapsing Quazi-monochromatic Waves

We select a particular form of the pulse by setting the second derivative $f''(x)$ as:

$$f''(x) = g(x) = C_1 e^{-\frac{x^2}{a^2}} \sin qx, \tag{28.29}$$

where C_1 is normalized coefficient defined by (see (28.21)):

$$C_1^2 = 3\varepsilon \left[a\sqrt{2\pi} \left(1 - e^{-\frac{q^2 a^2}{2}} \right) \right]^{-1} \tag{28.30}$$

Following Sects. 28.3 and 28.4, we will derive the pulse spectrum in far field region and the energy density in the center. By substituting (28.29) in (28.25) we get the spectral function of the fluence:

$$F_\omega(\vec{r}) = \frac{\sin^2\theta}{c^2 r^2} [g(\kappa)]^2 = \frac{a^2 C_1^2 \sin^2\theta}{16\pi c^2 r^2} \left[\exp\left\{ -\frac{(\omega - cq)^2 a^2}{4c^2} \right\} - \exp\left\{ -\frac{(\omega + cq)^2 a^2}{4c^2} \right\} \right]^2. \tag{28.31}$$

Then, using (28.40) and (28.41) we can determine the central frequency ω_0 and bandwidth $\delta\omega$ of pulse spectrum in the center. In the case of a quazi-monochromatic wave $aq \gg 1$ and we obtain:

$$\omega_0 = cq, \quad \delta\omega = \frac{c}{a}. \tag{28.32}$$

The energy density in the center is defined by formula (28.28): for the specific pulse (28.29) its maximum value ϵ_m will be achieved at $x = 0$ (i.e. $t = 0$) and equal to

$$\epsilon_m = \frac{\sqrt{2}}{3\pi^{3/2}} \frac{\varepsilon q^2}{a\left(1 - e^{-\frac{q^2 a^2}{2}}\right)} \tag{28.33}$$

In the case of quazi-monochromatic pulse with taking into account $aq \gg 1$ and (28.32) we can derive (anticipated to a factor) dependence of the maximum energy density ϵ_m in the center upon the full energy of collapsing pulse, its wavelength and bandwidth:

$$\epsilon_m = \frac{8\pi\sqrt{2\pi}}{3} \frac{\varepsilon}{\lambda_0^3} \frac{\delta\omega}{\omega_0}, \; \lambda_0 = \frac{2\pi}{\omega_0}, \; aq = \frac{\omega_0}{\delta\omega} \gg 1. \tag{28.34}$$

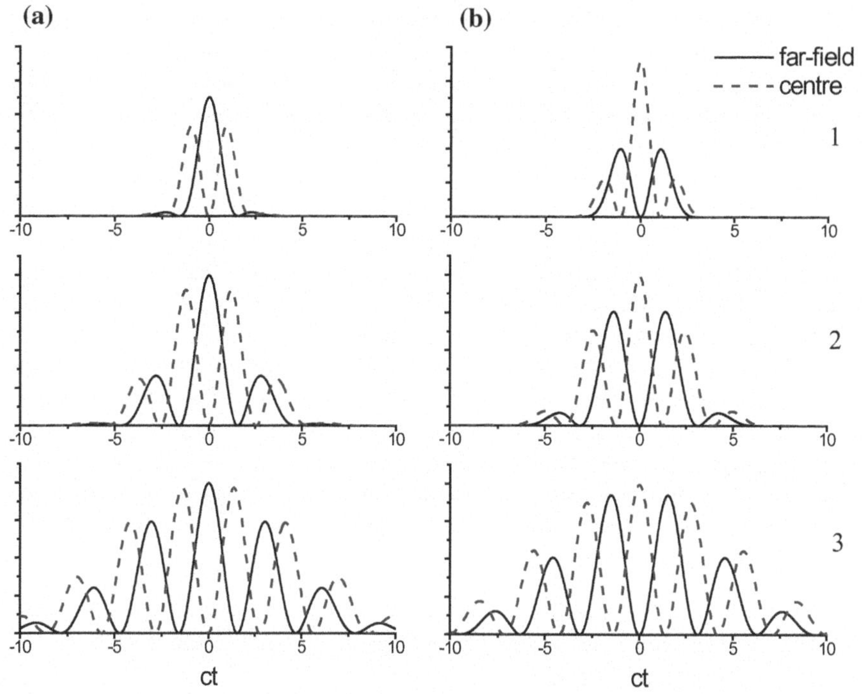

Fig. 28.1 Time dependence of the energy density $\epsilon(t)$ in the center (dashed blue line) and far-field region (solid black line) for different numbers of pulse cycles $N = aq$: $N = 2$ (1), $N = 4$ (2) and $N = 8$ (3). Figure 28.1a, b present even and odd functions $g(x)$, respectively (i.e. a swap between sin and cos functions in (28.29))

It's interesting to compare the time dependence of the energy density $\epsilon(t)$ in the center of collapsing pulse with that in far-field region. In the first case, this is a factor $[g'(ct)]^2$ (28.28), in the second—the square of the amplitude $[g(ct)]^2$. Figure 28.1 illustrates this fact by comparing the pulses with cycle numbers $N = aq$ in the range from 2 to 8, with "odd" and "even" denoting sin and cos in formula (28.29), respectively.

28.6 Discussion

Electromagnetic pulses that carry the limited energy and result from the exact solution of the Maxwell equations have been the subject of papers [2–7]. Above we considered one particular case to express the electromagnetic energy density in the focusing point in terms of the full energy and far-field spectrum (see (34)). We were able also to obtain the pulse shape in near and far-field regions.

Another possible application of similar exact solutions is a study of spatio-temporal couplings (STC) in ultrashort laser pulses. It is well known that a wavefront-based description of the pulses is inadequate in many problems of femto- and atto-physics and nonlinear optics [7–9]. Often the field should be calculated at a specific point defined by experimental conditions. We have to mention also the dependence of the pulse duration on a point in space [10]. More accurately to say that the spectral and time parameters of the field are different in its movement from one point in space to another. Both mentioned processes cannot be described with the help of Gaussian beams, parabolic wave equation or plane waves (despite the plane waves are the exact solution of the Maxwell equations). The reason is the infinite pulse energy that hinders an accurate theoretical modelling of experimental results.

On the other hand, the solution presented in this paper relates to the finite energy pulse and describes the corresponding STC effects. Thus, equations and formulas of this paper enable calculating a time dependence and field spectrum not only in the center of a collapsing pulse but at any distance from it.

We should mention also that the importance of exact solutions of electromagnetic wave equations for studies of the QED effects has been noticed in the work [4].

Annex: Formulas Useful for the Frequency and Bandwidth Analysis (See Sect. 28.5)

Let $g(t)$ be a real-bounded signal with the power equal (or proportional) to $[g(t)]^2$ and the total energy G equal to:

$$G = \int_{-\infty}^{\infty} \mathrm{d}t[g(t)]^2. \tag{28.35}$$

Then we can write the following expressions for the Fourier transform $g(\omega)$:

$$g(t) = \int_{-\infty}^{\infty} e^{-i\omega t} g(\omega)\mathrm{d}\omega, \quad g(\omega) = \frac{1}{2\pi}\int_{-\infty}^{\infty} e^{i\omega t} g(t)\mathrm{d}t, \tag{28.36}$$

$$g(-\omega) = g^*(\omega), |g(-\omega)|^2 = |g(\omega)|^2. \tag{28.37}$$

$$G = 2\pi \int_{-\infty}^{\infty} \mathrm{d}\omega |g(\omega)|^2 = 4\pi \int_{0}^{\infty} \mathrm{d}\omega |g(\omega)|^2 = \int_{0}^{\infty} \varepsilon(\omega)\mathrm{d}\omega, \tag{28.38}$$

where $\varepsilon(\omega)$ is the spectral density of the energy.

$$\varepsilon(\omega) = 4\pi |g(\omega)|^2. \tag{28.39}$$

The frequency ω_0 and the bandwidth $\delta\omega$ can be expressed in terms of the first and second moments of the spectral function $\varepsilon(\omega)$.

$$\omega_0 = G^{-1}\int_0^\infty \omega\varepsilon(\omega)\mathrm{d}\omega = G^{-1}4\pi\int_0^\infty \omega\mathrm{d}\omega|g(\omega)|^2, \tag{28.40}$$

$$(\delta\omega)^2 = G^{-1}\int_0^\infty (\omega-\omega_0)^2\varepsilon(\omega)\mathrm{d}\omega = G^{-1}4\pi. \tag{28.41}$$

References

1. A.N. Tikhonov, A.A. Samarsky, *The Equations of Mathematical Physics Textbook* (Moscow State University Press, 2004) (in Russian)
2. R.W. Ziolkowski, I.M. Besieris, A.M. Shaarawi, Localized wave representations of acoustic and electromagnetic radiation. Proc. IEEE **79**, 1371–1378 (1993)
3. J. Lekner, TM, TE and TEM beam modes: exact solutions and their problems. J. Opt. A: Pure Appl. Opt. **3**, 407–412 (2001)
4. A.M. Fedotov, K.Y. Korolev, M.V. Legkov, Exact analytical expression for the electromagnetic field in a focused laser beam or pulse, in *Proceedings of SPIE*, vol. 6726 (2007), p. 672613
5. A. April, Ultrashort, strongly focused laser pulses in free space, in *Coherence and Ultrashort Pulse Laser Emission* (InTech, 2010)
6. J. Lekner, *Theory of Electromagnetic Pulses* (Morgan & Claypool Publishers, San Rafael, 2018)
7. R.J. Ducharme, Constrained Bateman-Hillion solutions for Hermite-Gaussian beams (2014), arXiv:1412.1928
8. S. Akturk, X. Gu, P. Bowlan, R. Trebino, Spatio-temporal couplings in ultrashort laser pulses. J. Opt. **12**, 093001 (2010)
9. G. Pariente, V. Gallet, A. Borot, O. Gobert, F. Quéré, Space–time characterization of ultra-intense femtosecond laser beams. Nat. Photonics **10**, 547–553 (2016)
10. B. Alonso, J. Pérez-Vizcaíno, G. Mínguez-Vega, Í.J. Sola, Tailoring the spatio-temporal distribution of diffractive focused ultrashort pulses through pulse shaping. Opt. Express **26**, 10762–10772 (2018)

Author Index

A
Akhmedzhanov, T., 45
Albrecht, M., 93
Andreasson, J., 175
Andreev, A. V., 99
Ansaldi, G., 131
Antipenkov, R., 93
Antonov, V., 23, 45
Artyukov, I. A., 187

B
Bakule, P., 93
Balcou, Ph., 53, 79
Batysta, F., 93
Baynard, E., 53, 79
Beiersdorfer, P., 29
Bleiner, D., 3
Boge, R., 93
Boháček, K., 117
Bomme, C., 163
Bourassin-Bouchet, C., 163
Bravo, H., 11
Brown, C. R. D., 29
Brown, G. V., 29
Bulanov, S. V., 37
Burridge, D., 29

C
Cabas-Vidani, A., 3
Charles, R., 29
Chaulagain, U., 117
Chen, L. M., 109
Cubaynes, D., 163

D
Demailly, J., 53, 79
Deng, Y., 85
Dinh, T.-H., 37, 61
Dovillaire, G., 53, 79
Dyachkov, N. V., 187

E
Eisebitt, S., 131
Engel, D., 147
Esirkepov, T. Th., 37
Espinoza, S., 175

F
Faenov, A. Ya., 37
Feng, J., 109
Feshchenko, R. M., 181, 187
Finke, O., 93
Frassetto, F., 93
Friedrich, B., 131

G
Gajdoš, P., 103
Ganeev, R. A., 99
Garcia, G., 163
Gautier, J., 93, 155
Geloni, G., 131
Gharbi, A., 163
Giambruno, F., 93
Glöggler, L., 147
Goddet, J.-P., 155
Green, T., 93
Guggenmos, A., 155
Guilbaud, O., 53, 79, 163

M. Kozlová and J. Nejdl (eds.), *X-Ray Lasers 2018*, Springer Proceedings
in Physics 241, https://doi.org/10.1007/978-3-030-35453-4

Gu, M. F., 29
Gu, Y., 117

H
Hajima, R., 37
Hallmann, J., 131
Han, K. C., 45
Harms, F., 53, 79
Hasegawa, N., 37, 61, 67
Hatayama, M., 61
Hatfield, P., 29
Heathcote, R., 125
Helk, T., 155
Hell, N., 29
Higashiguchi, T., 61
Hill, M. P., 29
Hoarty, D., 29
Hobbs, L. M. R., 29
Hort, O., 93
Huang, K., 109
Hubka, Z., 93

I
Ichimaru, S., 61
Ishino, M., 37, 61

J
John, C., 67
Johnson, E., 79
Jonas, A., 147

K
Kalinin, N., 23
Kando, M., 37
Kawachi, T., 37, 61, 67
Kawase, K., 37
Kazamias, S., 53, 79
Khairulin, I., 45
Kiriyama, H., 37
Kishimoto, M., 67
Kitamura, T., 37
Kleineberg, U., 155
Klisnick, A., 163
Kocharovskaya, O., 45
Kolesnikov, A. O., 169
Kondo, K., 37
Kovalenko, A., 23
Kozlová, M., 73, 103, 117
Krůs, M., 73, 103
Kubo, N., 67
Kuznetsova, E., 45
Kuznezov, I., 3

L
Lamač, M., 117
Li, D. Z., 109
Li, R., 85
Li, W., 79
Li, Y. F., 109
Lolley, J., 11
Lucas, B., 53, 79
Lu, W., 131

M
Madsen, A., 131
Mai, D. D., 93
Mantouvalou, I., 147
Marcus, G., 85
Medvedev, N., 139
Menoni, C. S., 3, 11
Mikami, K., 37
Müller, R., 3
Murphy, C. D., 125

N
Nagai, R., 37
Namba, S., 67
Naylon, J., 93
Nefedova, V. E., 93
Nejdl, J., 93, 117
Neveu, O., 53, 79
Nilsen, J., 29
Nishikino, M., 37, 61, 67
Nishiuchi, M., 37
Noll, T., 131

O
Oliva, E., 17, 155

P
Panchenko, D., 29
Pandey, A. K., 53, 79, 163
Papagiannouli, E., 53, 79
Papagiannouli, I., 163
Patzke, G. R., 3
Pikuz, T. A., 37
Pirozhkov, A. S., 37, 169
Pittman, M., 53, 79
Poletto, L., 93

R
Radeonychev, Y., 45
Ragozin, E. N., 169
Reyne, S., 93
Richardson, M., 53, 79
Richter, S., 175
Rocca, J. J., 3, 11
Romanyuk, Y., 3
Ros, D., 53, 79
Rose, S., 29
Roth, T., 131
Rush, L., 3

S
Sagisaka, A., 37
Sakaki, H., 37
Sakaue, K., 61
Sanson, F., 53, 79
Sasaki, A., 37
Sasorov, P. V., 169
Saunders, A. M., 29
Scott, H. A., 29
Scully, M., 45
Sebban, S., 93, 155
Selwood, M. P., 125
Serkez, S., 131
Shatokhin, A. N., 169
Sinyakova, T., 163
Solis Meza, E., 11
Spielmann, C., 155
Stiel, H., 147, 155
Stremoukhov, S. Y., 99

T
Tallents, G. J., 11
Tan, J. H., 109
Tao, M. Z., 109
Ta-Phuoc, K., 117
Tissandier, F., 155
Tiwari, A., 3
Trottmann, M., 3
Tuitje, F., 155
Tümmler, J., 147

V
Vaganov, S., 23
Vančura, J., 117
Vinogradov, A. V., 187
Vishnyakov, E. A., 169

W
Wagenaars, E., 11
Wang, J. G., 109
Washio, M., 61
Weber, S. A., 117
Wilson, L., 29
Wilson, S. A., 11

Y
Yan, W., 117

Z
Zakopal, P., 73
Zeng, Z., 85
Zhang, X., 45
Zheng, Y., 85
Zhou, K., 131
Zürch, M., 155

The manufacturer's authorised representative in the EU is Springer Nature Customer Service Centre GmbH, Europaplatz 3, 69115 Heidelberg, Germany. If you have any concerns regarding our products, please contact ProductSafety@springernature.com

Printed and bound by CPI Group (UK) Ltd, Croydon, CR0 4YY
30/09/2026
02225646-0011